London Mathematical Society Lecture Note Series 80

Introduction to the Representation Theory of Compact and Locally Compact Groups

ALAIN ROBERT

CAMBRIDGE UNIVERSITY PRESS

LONDON MATHEMATICAL SOCIETY LECTURE NOTE SERIES

Managing Editor: Professor I.M. James,
Mathematical Institute, 24-29 St Giles, Oxford

1. General cohomology theory and K-theory, P.HILTON
4. Algebraic topology, J.F.ADAMS
5. Commutative algebra, J.T.KNIGHT
8. Integration and harmonic analysis on compact groups, R.E.EDWARDS
9. Elliptic functions and elliptic curves, P.DU VAL
10. Numerical ranges II, F.F.BONSALL & J.DUNCAN
11. New developments in topology, G.SEGAL (ed.)
12. Symposium on complex analysis, Canterbury, 1973, J.CLUNIE & W.K.HAYMAN (eds.)
13. Combinatorics: Proceedings of the British Combinatorial Conference 1973, T.P.McDONOUGH & V.C.MAVRON (eds.)
15. An introduction to topological groups, P.J.HIGGINS
16. Topics in finite groups, T.M.GAGEN
17. Differential germs and catastrophes, Th.BROCKER & L.LANDER
18. A geometric approach to homology theory, S.BUONCRISTIANO, C.P. BOURKE & B.J.SANDERSON
20. Sheaf theory, B.R.TENNISON
21. Automatic continuity of linear operators, A.M.SINCLAIR
23. Parallelisms of complete designs, P.J.CAMERON
24. The topology of Stiefel manifolds, I.M.JAMES
25. Lie groups and compact groups, J.F.PRICE
26. Transformation groups: Proceedings of the conference in the University of Newcastle-upon-Tyne, August 1976, C.KOSNIOWSKI
27. Skew field constructions, P.M.COHN
28. Brownian motion, Hardy spaces and bounded mean oscillations, K.E.PETERSEN
29. Pontryagin duality and the structure of locally compact Abelian groups, S.A.MORRIS
30. Interaction models, N.L.BIGGS
31. Continuous crossed products and type III von Neumann algebras, A.VAN DAELE
32. Uniform algebras and Jensen measures, T.W.GAMELIN
33. Permutation groups and combinatorial structures, N.L.BIGGS & A.T.WHITE
34. Representation theory of Lie groups, M.F. ATIYAH et al.
35. Trace ideals and their applications, B.SIMON
36. Homological group theory, C.T.C.WALL (ed.)
37. Partially ordered rings and semi-algebraic geometry, G.W.BRUMFIEL
38. Surveys in combinatorics, B.BOLLOBAS (ed.)
39. Affine sets and affine groups, D.G.NORTHCOTT
40. Introduction to Hp spaces, P.J.KOOSIS
41. Theory and applications of Hopf bifurcation, B.D.HASSARD, N.D.KAZARINOFF & Y-H.WAN
42. Topics in the theory of group presentations, D.L.JOHNSON
43. Graphs, codes and designs, P.J.CAMERON & J.H.VAN LINT
44. Z/2-homotopy theory, M.C.CRABB
45. Recursion theory: its generalisations and applications, F.R.DRAKE & S.S.WAINER (eds.)
46. p-adic analysis: a short course on recent work, N.KOBLITZ
47. Coding the Universe, A.BELLER, R.JENSEN & P.WELCH
48. Low-dimensional topology, R.BROWN & T.L.THICKSTUN (eds.)

49. Finite geometries and designs, P.CAMERON, J.W.P.HIRSCHFELD & D.R.HUGHES (eds.)
50. Commutator calculus and groups of homotopy classes, H.J.BAUES
51. Synthetic differential geometry, A.KOCK
52. Combinatorics, H.N.V.TEMPERLEY (ed.)
53. Singularity theory, V.I.ARNOLD
54. Markov processes and related problems of analysis, E.B.DYNKIN
55. Ordered permutation groups, A.M.W.GLASS
56. Journées arithmétiques 1980, J.V.ARMITAGE (ed.)
57. Techniques of geometric topology, R.A.FENN
58. Singularities of smooth functions and maps, J.MARTINET
59. Applicable differential geometry, F.A.E.PIRANI & M.CRAMPIN
60. Integrable systems, S.P.NOVIKOV et al.
61. The core model, A.DODD
62. Economics for mathematicians, J.W.S.CASSELS
63. Continuous semigroups in Banach algebras, A.M.SINCLAIR
64. Basic concepts of enriched category theory, G.M.KELLY
65. Several complex variables and complex manifolds I, M.J.FIELD
66. Several complex variables and complex manifolds II, M.J.FIELD
67. Classification problems in ergodic theory, W.PARRY & S.TUNCEL
68. Complex algebraic surfaces, A.BEAUVILLE
69. Representation theory, I.M.GELFAND et. al.
70. Stochastic differential equations on manifolds, K.D.ELWORTHY
71. Groups - St Andrews 1981, C.M.CAMPBELL & E.F.ROBERTSON (eds.)
72. Commutative algebra: Durham 1981, R.Y.SHARP (ed.)
73. Riemann surfaces: a view toward several complex variables, A.T.HUCKLEBERRY
74. Symmetric designs: an algebraic approach, E.S.LANDER
75. New geometric splittings of classical knots (algebraic knots), L.SIEBENMANN & F.BONAHON
76. Linear differential operators, H.O.CORDES
77. Isolated singular points on complete intersections, E.J.N.LOOIJENGA
78. A primer on Riemann surfaces, A.F.BEARDON
79. Probability, statistics and analysis, J.F.C.KINGMAN & G.E.H.REUTER (eds.)
80. Introduction to the representation theory of compact and locally compact groups, A.ROBERT
81. Skew fields, P.K.DRAXL

London Mathematical Society Lecture Note Series. 80

Introduction to the Representation Theory of Compact and Locally Compact Groups

ALAIN ROBERT

Professor of Mathematics
University of Neuchatel

CAMBRIDGE UNIVERSITY PRESS
Cambridge
London New York New Rochelle
Melbourne Sydney

Published by the Press Syndicate of the University of Cambridge
The Pitt Building, Trumpington Street, Cambridge CB2 1RP
32 East 57th Street, New York, NY 10022, USA
296 Beaconsfield Parade, Middle Park, Melbourne 3206, Australia

First published 1983

Printed in Great Britain at the University Press, Cambridge

Library of Congress catalogue number: 82-19730

British Library cataloguing in publication data

Robert, Alain
Introduction to the representation theory of compact and locally compact groups. - (London Mathematical Society lecture note series, ISSN 0076-0552; 80)
1. Groups, Theory
I. Title II. Series
512'.22 QA171

ISBN 0 521 28975 0

CONTENTS

Part II : Representations of locally compact groups

FOREWORD

These are notes from a graduate course given in Lausanne (Switzerland) during the winter term 1978-79 (Convention romande des enseignements de 3e cycle en mathématiques).

This term was devoted to a self-contained approach to representation theory for locally compact groups, using only *integral methods*. The sole prerequisite was a basic familiarity with the theory for finite groups (e.g. as contained in the first chapter of Serre 1967). For didactic reasons, I spent the first half of the term discussing compact groups, trying to be more elementary and more complete in this part. In particular, I have given several proofs of the main results. For example, the Peter-Weyl theorem is proved first with the use of the Stone-Weierstrass approximation theorem (p.33) and then without it (p.36). The "finiteness theorem" (irreducible representations of compact groups are finite dimensional) is proved first for Banach (or barrelled) spaces ((5.8)p.46), then in the general case (quasi-complete spaces) ((7.9)p.69) and finally in a more elementary fashion for Hilbert spaces ((8.5)p.81) . Thus I hope that readers with various backgrounds will be able to benefit from these notes.

My way of introducing the subject has forced me to repeat some definitions in the second part where I gradually assume more from my reader (this is particularly so as far as measure theory is concerned). This part in no way claims to be complete and only has an introductory purpose. I consider that the existing books on the subject more than prove that complete treatises on the subject are heavy going...

Finally, I should mention that the representation theory for Lie groups - starting with the rotation group $SO_3(\mathbb{R})$ and compact Lie groups - was considered in subsequent graduate courses, but has not been included in these notes since *differential methods* in the representation theory of Lie groups are probably more readily available in recent texts.

My presentation of the subject has certainly been influenced by R. Godement whose courses introduced me to this field. I would like to thank him here.

I would also like to thank Sylvie Griener who read the manuscript and helped me to detect various inaccuracies, and my wife Ann for hints on language.

July 1982

Alain Robert
Institut de Mathématiques
Université de Neuchatel
Chantemerle 20
CH-2000 NEUCHATEL 7
(Switzerland)

Conventional notations and terminology

$\mathbb{N} = \{0,1,2,\ldots\}$, $\mathbb{Z}$, $\mathbb{Q}$, $\mathbb{R}$, $\mathbb{C}$ fundamental numerical sets

$\mathbb{H}$ real quaternions (skew-field, $\dim_{\mathbb{R}}(\mathbb{H}) = 4$, $\mathbb{R}$-basis $1,i,j,k$)

$\mathbb{F}_q$ finite field with q elements ($q = p^f$ for some prime p)

$\mathbb{Q}_p$ field of p-adic numbers, $\mathbb{Q}_p \supset \mathbb{Z}_p$ ring of p-adic integers

$A^\times$ group of units in a ring A ($k^\times = k - \{0\}$ if k is a field)

$\mathbb{R}_+^\times$ multiplicative group of positive real numbers $x > 0$

(= neutral connected component of $\mathbb{R}^\times$)

$\emptyset$ empty set, Card(X) = number of elements of X

countable: finite or denumerable (equipotent to some part of $\mathbb{N}$)

$E \hookrightarrow F$ injective (one-one into) map

χ_A characteristic function of a subset $A \subset X$: = 1 on A, = 0 outside A

$f|_A$ restriction of a mapping to a subset $A \subset X$

$C(X,E)$ space of continuous maps $f : X \to E$

$C(X) = C(X,\mathbb{C})$, $C_{\mathbb{R}}(X) = C(X,\mathbb{R})$ continuous numerical functions on X

$C_c(X)$ subspace of $C(X)$ consisting of functions with compact support

Sup ess $|f|$: smallest M with $|f| \leqslant M$ nearly everywhere (on a measured space)

$\check{f}(x) = f(x^{-1})$ symmetric function on a group G

$f^*(s) = \overline{f(s^{-1})}$ on a unimodular group (cf. (14.2) p.133 in general)

scalar products are always linear in the *second* factor

$$a(x \mid y) = (x \mid ay) = (\bar{a}x \mid y)$$

normal subgroup = invariant subgroup (= distinguished subgroup)

commutative = abelian (= Abelian!)

in sec. 20-21, separable group means locally compact group admitting a countable basis for the open sets (hence has a countable dense subset)

$1_n = \mathrm{id}_n = \mathrm{id}.$ unit matrix (in dim n)

${}^tg$, ${}^tA$ transpose of a matrix ($A^* = {}^t\bar{A}$, $\check{g} = {}^tg^{-1}$ contragredient)

Π_n vector space of (complex) polynomials in z and degree $\leqslant n$ (p.102)

X^G set of fixed points in a group action of G on X

Hausdorff space = T_2-space (= separated space)

Part I :

Representations of compact groups

1 COMPACT GROUPS AND HAAR MEASURES

Before starting representation theory, it is certainly appropriate to start with a review of examples of compact groups.

EXAMPLES

Some matrix groups

Let $O_n(\mathbb{R})$ denote the group of real $n\times n$ matrices which preserve the standard quadratic form $x_1^2 + x_2^2 + \ldots + x_n^2$. Elements of this group are real $n\times n$ matrices g satisfying the relation ${}^t g\, g = 1_n$ (the columns of the matrix g must constitute an orthonormal basis of $\mathbb{R}^n$). This group is the *orthogonal group* : it is a compact subgroup of the general linear group $Gl_n(\mathbb{R})$ in n real variables. Since the relation ${}^t g\, g = 1_n$ implies $(\det g)^2 = 1$ hence $\det g = \pm 1$ and both cases occur in $O_n(\mathbb{R})$, we see that this group is not connected. Its index two subgroup

$$SO_n(\mathbb{R}) = O_n(\mathbb{R}) \cap Sl_n(\mathbb{R}) \quad \textit{special orthogonal group}$$

is known to be connected (Chevalley 1946, Dieudonné 1970 (16.11.7) p.68). The first non-trivial group in this series is the *circle group*

$$\mathbb{R}/\mathbb{Z} \longrightarrow SO_2(\mathbb{R}) \quad \text{(isomorphism)}$$

$$t \bmod \mathbb{Z} \longmapsto e^{2\pi i t} = a + ib \longmapsto \begin{pmatrix} a & b \\ -b & a \end{pmatrix} \quad (a^2 + b^2 = 1).$$

The next one is the *rotation group* $SO_3(\mathbb{R})$ (we shall study it in detail). For $n \geqslant 3$, the (special) orthogonal groups $SO_n(\mathbb{R})$ are not commutative.

We can use the complex field $\mathbb{C}$ (instead of $\mathbb{R}$) and consider the hermitian form $\bar{z}_1 z_1 + \bar{z}_2 z_2 + \ldots + \bar{z}_n z_n$ on $\mathbb{C}^n$, thus defining *unitary transformations* g as complex $n\times n$ matrices satisfying $g^* g = 1_n$ (recall that $g^* = {}^t\bar{g}$). The *unitary group* $U_n(\mathbb{C})$ is a compact connected (loc. cit.) subgroup of $Gl_n(\mathbb{C})$. The first such group is the circle group $U_1(\mathbb{C})$ (identified with the multiplicative group of complex numbers of modulus 1). We shall see later that the quotient of $U_2(\mathbb{C})$ by its center is (isomorphic to) $SO_3(\mathbb{R})$. Quite generally, the circle group $U_1(\mathbb{C})$ can be embedded *diagonally* into $U_n(\mathbb{C})$, the image of this embedding $U_1(\mathbb{C}) \hookrightarrow U_n(\mathbb{C})$

being the center of the unitary group $U_n(\mathbb{C})$. Thus the center of $U_n(\mathbb{C})$ is connected (but the center of $O_n(\mathbb{R})$ is finite). Imposing furthermore the determinant 1 condition, we define the *special unitary groups*

$$SU_n(\mathbb{C}) = U_n(\mathbb{C}) \cap Sl_n(\mathbb{C}) .$$

(The condition $g^* g = 1_n$ in $U_n(\mathbb{C})$ implies $|\det g|^2 = 1$ and all complex numbers having modulus 1 are determinants of elements of $U_n(\mathbb{C})$.) The center of $SU_n(\mathbb{C})$ consists of the scalar matrices of determinant 1 : it is a cyclic group of order n isomorphic to the subgroup of nth roots of 1 (in $\mathbb{C}^\times$ or $U_1(\mathbb{C})$).

Similar considerations hold over the field $\mathbb{H}$ of real quaternions (the involution $q \mapsto \bar{q}$ being the quaternionic conjugation) with respect to the real bilinear form $\bar{q}_1 q_1 + \bar{q}_2 q_2 + \dots + \bar{q}_n q_n$. But since this field $\mathbb{H}$ is not commutative, some care has to be taken with respect to the representation of $\mathbb{H}$-linear mappings from $\mathbb{H}^n$ into itself by quaternionic $n \times n$ matrices (one should say *left* $\mathbb{H}$-linear mappings to be quite precise). Thus one can construct compact connected groups $U_n(\mathbb{H}) \supset SU_n(\mathbb{H})$, which are also called *symplectic* groups.

The three *series* $SO_n(\mathbb{R})$ $(n \geqslant 3)$, $SU_n(\mathbb{C})$ $(n \geqslant 2)$ and $SU_n(\mathbb{H})$ $(n \geqslant 1)$ are the *classical groups*. Together with five *exceptional* groups, they exhaust the list of compact connected "simple" groups (more precisely, their center is finite and the quotient by their center is simple).

Some connected groups (not Lie groups)

The simplest example of a compact connected group which is not a Lie group is certainly the group $G = (\mathbb{R}/\mathbb{Z})^{\mathbb{N}}$ (infinite product of circle groups). This group is commutative and each neighbourhood of its neutral element 0 contains a subgroup

$$G_n = \{0\} \times (\mathbb{R}/\mathbb{Z})^{[n,\infty[}$$

(this follows immediately from the definition of the product topology). Thus G contains *arbitrarily small subgroups*. More generally, let $(G_i)_I$ be a family of (non-trivial) compact connected groups. The product $G = \prod G_i$ is also a compact and connected group. Since a topological group is *metrizable* exactly when there is a countable fundamental system of neighbourhoods of its neutral element, we see that such products are metrizable precisely when the family I is countable. For compact groups, the following properties are equivalent (cf. also (5.11))

i) G is metrizable,

ii) G has a countable basis for open sets.

They imply

iii) G is separable (there is a countable family which is everywhere dense in G).

Totally discontinuous groups

In any topological group G, the connected component of the neutral element G^o is a closed and normal subgroup. When $G^o = \{e\}$, the only connected subsets of G are the points (and the empty set!) and G is *totally discontinuous*. In general, G/G^o will be totally discontinuous. A locally compact space which is totally discontinuous has the following property: each point has a fundamental system of open and closed neighbourhoods (Bourbaki 1971, TG II cor. of prop.6 p.32). The simplest example of totally discontinuous compact group is the *Cantor group*

$$G = (\mathbb{Z}/2\mathbb{Z})^{\mathbb{N}} .$$

Its elements are the infinite sequences $a = (a_n)$ with $a_n = 0$ or 1 $(n \in \mathbb{N})$. The topological space underlying this group is usually obtained by removing successively from the unit interval $I = [0,1]$, its middle third $]1/3,2/3[$, then removing from each remaining interval its (open) third, etc... (cf. picture below). The intersection of this decreasing family of compact sets is the Cantor set. The topology induced by the real line coïncides with the product topology when elements of this set are represented in the dyadic system (as usual).

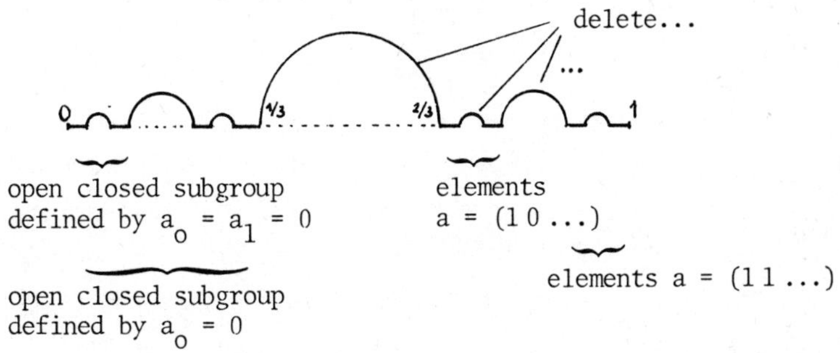

The Cantor group is commutative and is not the most interesting totally discontinuous group. More sophisticated examples of totally discontinuous groups occur naturally in two contexts.

p-adic groups. The topological ring of p-adic integers $\mathbb{Z}_p$ can be defined either as (here, p is any prime number)

completion $\widehat{\mathbb{Z}_{(p)}}$ of the local ring $\mathbb{Z}_{(p)} \subset \mathbb{Q}$ (consisting of fractions a/b with b not divisible by p),

or as

inverse limit $\varprojlim_n \mathbb{Z}/p^n\mathbb{Z}$ (with respect to the canonical homomorphisms $\mathbb{Z}/p^{n+1}\mathbb{Z} \to \mathbb{Z}/p^n\mathbb{Z}$ of finite rings).

The second definition makes obvious the fact that $\mathbb{Z}_p$ is compact. The additive group $\mathbb{Z}_p$ and the multiplicative group $\mathbb{Z}_p^\times$ are compact commutative groups. The groups

$$\mathrm{Gl}_n(\mathbb{Z}_p) : \text{p-adic } n \times n \text{ matrices } g \text{ with } \det g \in \mathbb{Z}_p^\times$$

are good examples of totally discontinuous compact groups (for $n \geqslant 2$, these groups are not commutative).

Galois groups. Let k be any field, k_s denoting a *separable closure* of k (in some algebraic closure $\bar{k}$ of k). The group

$$G = \mathrm{Gal}(k_s/k) = \mathrm{Aut}_k(k_s)$$

is a compact topological group with respect to the *Krull topology*. Recall that this topology on G is defined by taking for fundamental system of neighbourhoods of the neutral element of G (the identity of k_s) the subgroups $G' = \mathrm{Gal}(k_s/k')$ for all *finite* (separable) subextensions k' of k. These subgroups G' are both open and closed so that G is totally discontinuous. Moreover, with this topology, there is a Galois correspondence (inclusion reversing)

$$\left\{\begin{matrix}\text{closed subgroups}\\ \text{of } G\end{matrix}\right\} \longleftrightarrow \left\{\begin{matrix}\text{intermediate extensions}\\ \text{between } k \text{ and } k'\end{matrix}\right\}$$

$$H = \mathrm{Gal}(k_s/L) \longleftrightarrow L = \text{Fixed field under } H \quad .$$

Compact totally discontinuous groups are always pro-finite groups.

HAAR MEASURE

The main analytical tool in the study of compact groups is the Haar integral. We shall construct this integral (or Radon measure) for continuous functions only, assuming that the reader is familiar with the extension procedure to Borel functions and sets (and eventually to measurable sets). However, very little of the abstract integration theory is needed : L^p spaces can be defined abstractly as suitable completions of the space of continuous fonctions ($1 \leqslant p < \infty$, $p = 1$ and 2 being the most important cases). Negligible sets can be defined as those sets N which have the following property : for each $\varepsilon > 0$, there is a continuous positive function $f = f_\varepsilon$ which has a restriction to N, $f_N \geqslant 1$ and integral smaller than ε. Here is the main statement.

Theorem. Let G be a compact group. Then, there is a unique linear form

$$m : C(G) \longrightarrow \mathbb{C} \quad (C(G) : \text{space of continuous maps } G \longrightarrow \mathbb{C})$$

having the properties

1. $m(f) \geqslant 0$ for $f \geqslant 0$ (m is *positive*) ,
2. $m(1) = 1$ (m is *normalized*) ,
3. $m({}_sf) = m(f)$ for ${}_sf(g) = f(s^{-1}g)$ $(s , g \in G)$
   (m is *left invariant*) .

Moreover, this linear form m is also *right invariant*

4. $m(f_s) = m(f)$ for $f_s(g) = f(gs)$ $(s , g \in G)$.

The proof of this theorem can be based on the following two classical results.

a) *(Ascoli's theorem) Let X be a compact topological space, E a Banach space and Φ a subset of $C_E(X) = C(X;E)$ (Banach space of continuous mappings $X \longrightarrow E$ with the uniform norm). Then Φ is relatively compact (i.e. has a compact closure in $C_E(X)$) if and only if Φ is equicontinuous and all sets $\Phi(x) = \{f(x) : f \in \Phi\}$ $(x \in X)$ are relatively compact in E.*

(cf. Dieudonné 1960, p.137 for the metric case, or Rudin 1973, p.369 .)

b) *(Kakutani's fixed point theorem) Let E be a Banach space (or any locally convex topological vector space), K a convex $\neq \emptyset$ compact subset of E, and G a compact group acting linearly on E. If the action $\lambda : G \longrightarrow Gl(E)$ leaves K invariant*

$\lambda(g)K \subset K$ $(g \in G)$ and $\Phi = \lambda(G)$ is equicontinuous, then there is a fixed point of G in K.

(cf. Rudin 1973, p.120 .)

Here is a construction of the measure m. For $f \in C(G)$ (i.e. f is a continuous complex valued function on G) we denote by C_f the convex hull of all left translates ${}_sf$ $(s \in G)$ of f. Thus, the elements of C_f are the functions which are finite sums of the form

$$\sum_{\text{finite}} a_i\, f(s_i x) \quad \text{with} \quad a_i > 0 \text{ and } \Sigma a_i = 1 \; .$$

Clearly if $g \in C_f$

$$\|g\| = \operatorname*{Max}_G |g(x)| \leqslant \operatorname*{Max}_G |f(x)| = \|f\| \; .$$

In particular, all sets $C_f(x) = \{g(x) : g \in C_f\}$ are bounded and relatively compact in $\mathbb{C}$. On the other hand, since G is compact, f is *uniformly* continuous on G : for each $\varepsilon > 0$, there exists a neighbourhood $V = V_\varepsilon$ of the neutral element $e \in G$ with the property

$$y^{-1}x \in V \implies |f(y) - f(x)| < \varepsilon \; .$$

Since $(s^{-1}y)^{-1}s^{-1}x = y^{-1}ss^{-1}x = y^{-1}x$, we shall also have

$$|{}_sf(y) - {}_sf(x)| < \varepsilon \quad \text{as soon as} \quad y^{-1}x \in V \; .$$

Making convex combinations of these, we also infer

$$|g(y) - g(x)| < \varepsilon \quad \text{as soon as} \quad y^{-1}x \in V \quad (\text{all } g \in C_f).$$

This proves that the set C_f is uniformly equicontinuous. By Ascoli's theorem, we conclude that C_f is relatively compact in C(G). Let K_f denote the closure of C_f in C(G): this is a compact convex set. The compact group G acts by left translations (isometrically) on C(G) and leaves C_f hence also $K_f = \bar{C}_f$ invariant. Kakutani's theorem asserts then that there is a fixed point of this action of G in K_f. Such a fixed point is a constant function

$${}_sg = g \quad (s \in G) \implies g(s^{-1}) = {}_sg(e) = g(e) = c \quad (s \in G).$$

Such a constant has the property of being approximated by elements of C_f i.e. convex combinations of left translates of f :
for each $\varepsilon > 0$, there are finitely many $s_i \in G$ and $a_i > 0$ with

$$\Sigma a_i = 1 \quad \text{and} \quad |c - \Sigma a_i f(s_i x)| < \varepsilon \quad (x \in G) \qquad (1).$$

Let us show that there is only one constant function in K_f. For this, we start the same construction with *right* translations of f (we can apply the preceding construction to the opposite group G' of G, or the function $f'(x) = f(x^{-1})$ ...), obtaining a relatively compact convex set C'_f with compact convex closure K'_f containing a constant function c'. It will be enough to show c = c' (all constants c in K_f must be equal to *one chosen constant* c' of K'_f, and conversely !). There is certainly a finite combination of right translates which is close to c'

$$|c' - \sum b_j f(xt_j)| < \varepsilon \quad \text{(for some } t_j \in G,\ b_j > 0 \text{ with } \sum b_j = 1).$$

Let us multiply this equality by a_i and put $x = s_i$

$$|c'a_i - \sum a_i b_j f(s_i t_j)| < \varepsilon a_i \tag{2}$$

Summing over i, we obtain

$$|c'\sum a_i - \sum_{i,j} a_i b_j f(s_i t_j)| < \varepsilon \sum a_i = \varepsilon \tag{3}$$

Operating symmetrically on (1) (multiplying by b_j, putting $x = t_j$ and summing over j), we find

$$|c' - \sum_{i,j} a_i b_j f(s_i t_j)| < \varepsilon \tag{4}$$

Adding (or subtracting !) (3) and (4), we conclude $|c - c'| < 2\varepsilon$! Since ε can be taken arbitrarily small, we must have c = c' (uniqueness). From now on, *the* constant c in K_f will be denoted by m(f) : it is the only constant function which can be approximated arbitrarily close with convex combinations of left (or right) translates of f. The following properties are obvious

$m(1) = 1 \qquad (K_f = \{1\} \text{ if } f = 1)$,

$m(f) \geq 0$ if $f \geq 0$,

$m(af) = a\,m(f)$ if a is any complex number $(K_{af} = aK_f)$,

$m({}_sf) = m(f) = m(f_s)$ (by uniqueness) .

Our proof will be complete if we show that m is *additive* (hence *linear*). Let us take $f, g \in C(G)$ and start with (1) above with c = m(f). Put moreover

$$h(x) = \sum a_i g(s_i x) \quad .$$

Since $h \in C_g$, we certainly have $C_h \subset C_g$ whence $K_h \subset K_g$. But the set K_g contains only one constant : m(h) = m(g) . We can write

$$\left| m(h) - \sum b_j h(t_j x) \right| < \varepsilon$$

for finitely many suitable $t_j \in G$, $b_j > 0$ and $\sum b_j = 1$. Using the definition of h and m(h) = m(g), we find

$$\left| m(g) - \sum_{i,j} a_i b_j g(s_i t_j x) \right| < \varepsilon \tag{5}$$

But multiplying (1) by b_j, replacing x by $t_j x$ in it and summing over j we find

$$\left| m(f) - \sum_{i,j} a_i b_j f(s_i t_j x) \right| < \varepsilon \tag{6}$$

Adding (5) and (6), we find

$$\left| m(f) + m(g) - \sum_{i,j} a_i b_j (f+g)(s_i t_j x) \right| < 2\varepsilon .$$

Hence the constant m(f) + m(g) is in K_{f+g} : but the only constant of this compact convex set is m(f + g). q.e.d.

This measure m on G is the (normalized) *Haar measure* of the compact group G. Instead of m(f), we shall often write

$$\int_G f(x)\, dm(x) \quad \text{or even, more simply} \quad \int f(x)\, dx .$$

This notation will also be used for the (regular Borel) extended measure and integrable functions $f \in L^1(G) = L^1(G,m)$. It is countably additive. For example, the invariance under translations implies that points of G have same measure, and since m(G) = 1, we infer

$$m(\{e\}) > 0 \implies G \text{ finite group} .$$

In this case, the (normalized) Haar measure of G is simply given by

$$m(\{e\}) = 1/n \quad (n = \mathrm{Card}\, G) ,$$

$$m(f) = \frac{1}{n} \sum_{x \in G} f(x) \qquad (f \text{ in the } \textit{group algebra} \text{ of } G) .$$

In the opposite case, m({e}) = 0, all points have measure 0 and G cannot be countable (by countable additivity of m, m(N) = 0 for all countable sets in G : countable sets of G are negligible).

The measure of a subset of G is (by definition) the measure (or integral) of the characteristic function of this set. For measurable sets (or more simply Borel sets), m(A) is at the same time the supremum of the measures m(K) of the compact subsets of A, or the infimum of the measures m(U) of the open neighbourhoods U of A.

EXERCISES

1. Let G be a compact group and m its Haar measure. Show that $m: C_{\mathbb{R}}(G) \longrightarrow \mathbb{R}$ is continuous ($C_{\mathbb{R}}(G)$ is the Banach space with respect to the sup norm). Hint: use $-\|f\| \leqslant f(x) \leqslant \|f\|$ $(f \in C_{\mathbb{R}}(G),\ x \in G)$.

2. Take for G the Cantor group. What is the measure of the subgroup defined by $a_o = 0$? Same question for the subgroups defined by $a_i = 0$ for $i \leqslant n$. Deduce from the preceding observations an expression for the Haar integral of a locally constant function. This gives a good "feeling" for the Haar integral of any continuous function (any such function is uniformly continuous hence uniformly approximable by locally constant functions and ex.1 can be applied).

Let now $I = [0,1]$ be the unit interval and consider G as embedded (topologically and metrically) in I. Show that G is a Lebesgue negligible set in I. The measure m on I (with support G) is not absolutely continuous with respect to Lebesgue measure (m is singular with respect to Lebesgue measure). The function f on I defined by $f(x) = m([0,x])$ is increasing and continuous. It is continuously differentiable in $I - G$ (remember that G is negligible) with $f'(x) = 0$ for $x \notin G$. However, f is not constant in I ! (One should recall that if f is differentiable outside a <u>countable</u> subset A of I with $f'(x) = 0$ for all $x \notin A$, then f is constant. One cannot replace the assumption A countable by A negligible.)

3. Let p be any prime number and consider the Haar measure m of the additive (compact) group $\mathbb{Z}_p$. What is the measure of the subgroups $p^n\mathbb{Z}_p$?

Show that the restriction of m to $\mathbb{Z}_p^{\times}$ is (proportional to) a Haar measure of this multiplicative group. Hint: observe that $1 + p\mathbb{Z}_p$ is a subgroup of index $p-1$ in the multiplicative group $\mathbb{Z}_p^{\times}$. Observe then that $1 + p^n\mathbb{Z}_p$ is a subgroup of index p^{n-1} of the multiplicative group $1 + p\mathbb{Z}_p$ (similarity with: $p^n\mathbb{Z}_p$ is a subgroup of index p^{n-1} in the additive group $p\mathbb{Z}_p$, and a subgroup of index p^n in $\mathbb{Z}_p$).

More precisely, show that the restriction of $p/(p-1)\cdot m$ to $\mathbb{Z}_p^{\times}$ is the normalized Haar measure of this multiplicative group.

4. Show that the Haar measure of $SO_2(\mathbb{R})$ is given by (the absolute value of) the differential form

$$db/(2\pi a) = -da/(2\pi b)$$

in terms of the parameters a and b of $x = \begin{pmatrix} a & b \\ -b & a \end{pmatrix}$ $(a^2 + b^2 = 1)$.

Show similarly that the Haar measure of $U_1(\mathbb{C})$ is given by (the absolute value of) the differential form

$dz/(2\pi iz)$ (this is a real 1-form)

in term of the parameter $z \in \mathbb{C}^{\times}$, $|z| = 1$.

(Remark. Although very few books insist on this point, a measure is canonically associated to the absolute value of a real exterior differential form (on an oriented manifold). For example, Lebesgue measure dxdy in $\mathbb{R}^2$ is the absolute value of the exterior 2-form $dx \wedge dy$:

$$dxdy = |dx \wedge dy| .$$

The relation $dy \wedge dx = - dx \wedge dy$ is thus compatible with Fubini's theorem: $dxdy = dydx$.)

5. Let m be the (normalized) Haar measure of a compact group G. Show that $m(\check{f}) = m(f)$ (remember that $\check{f}(x) = f(x^{-1})$) for $f \in C(G)$ or $f \in L^1(G)$. This equality is usually written

$$\int_G f(x)\,dx = \int_G f(x^{-1})\,dx .$$

(Observe that $f \mapsto m(\check{f})$ is a Haar measure on G and use the uniqueness part of the theorem on Haar measures.)

2 REPRESENTATIONS, GENERAL CONSTRUCTIONS

If E is a (complex) Banach space, we denote by Gl(E) the group of continuous isomorphisms of E onto itself (if u is linear, continuous and bijective, u^{-1} will automatically be continuous, but we can require more simply u and u^{-1} to be continuous... and we *should* do it if E were a more general *locally convex topological vector space...*).

A <u>representation</u> π of a compact group G in E is a homomorphism

$$\pi : G \longrightarrow Gl(E)$$

for which all maps

$$G \longrightarrow E , \quad s \longmapsto \pi(s)\, v \qquad\qquad (v \in E)$$

are continuous. (I sincerely hope that the choice of Greek letter π for representations will not induce any confusion with the number $\pi = 3.14...$ even when both appear, e.g. for representations of rotation groups, where angles have to be introduced !)

The space $E = E_\pi$ in which the representation takes place is called <u>representation space</u> of π.

A representation π of a group G in a vector space E canonically defines an <u>action</u> (also denoted π)

$$\pi : G \times E \longrightarrow E$$
$$(s,v) \longmapsto \pi(s)\, v \quad .$$

The definition requires that this action is *separately continuous.* The action is then automatically *globally continuous* (this implication holds quite generally for barrelled spaces, as is explained in Bourbaki 1963 INT. Chap.VIII §2 Prop.1 p.130) .

One <u>should require</u> *global continuity* of the action in the definition for general locally convex topological vector spaces.

We say that the representation π is <u>unitary</u> when E = H is a Hilbert space and each operator $\pi(s)$ ($s \in G$) is a unitary operator (thus each $\pi(s)$ must be *isometric and surjective*). Thus π is unitary when E = H is a Hilbert space and

$$\pi(s)^* = \pi(s)^{-1} = \pi(s^{-1}) \qquad (s \in G) .$$

The representation π of G in E is said to be <u>irreducible</u> when E and $\{0\}$ are distinct and are the only two closed invariant subspaces under all operators $\pi(s)$ $(s \in G)$ (*topological irreducibility*). Two representations π and π' of a same group G are called <u>equivalent</u> when the two spaces over which they act are "G-*isomorphic*", i.e. when there exists a (continuous) isomorphism $A : E \longrightarrow E'$ of their respective spaces with

$$A\,(\pi(s)\,v) = \pi'(s)\,A\,v \qquad (s \in G,\ v \in E) .$$

More generally, continuous linear operators $A : E \longrightarrow E'$ satisfying all commutation relations $A\,\pi(s) = \pi'(s)\,A$ $(s \in G)$ are called <u>intertwining operators</u> or G-<u>morphisms</u> (from π to π') and their set is a vector space denoted either by

$$\mathrm{Hom}_G(E,E') \quad \text{or by} \quad \mathrm{Hom}(\pi, \pi') .$$

The following two propositions are relatively elementary. The first one does not even use compactness of G whereas the second one uses the Haar measure of G for *averaging* over G (the reader who has some familiarity with the representation theory of finite groups will have no difficulty in recognizing how the algebraic proof has been adapted to our analytical context).

(2.1) <u>Proposition</u>. <u>Let</u> π <u>be a unitary representation of</u> G <u>in the Hilbert space</u> H. <u>If</u> H_1 <u>is an invariant subspace of</u> H (<u>with respect to all operators</u> $\pi(s)$, $s \in G$), <u>then the orthogonal</u> $H_2 = H_1^{\perp}$ <u>of</u> H_1 <u>in</u> H <u>is also invariant</u>.

<u>Proof</u>. We have to show that if $v \in H$ is orthogonal to H_1, then all $\pi(s)v$ are also orthogonal to H_1 $(s \in G)$. This is obvious since for any $x \in H_1$ we can write

$$(x \mid \pi(s)v) = (\pi(s)^*x \mid v) = (\pi(s^{-1})x \mid v) = 0$$

by assumption ($\pi(s^{-1})x$ also lies in H_1). q.e.d.

(2.2) <u>Proposition</u>. <u>Let</u> π <u>be a representation of a compact group</u> G <u>in a Hilbert space</u> H. <u>Then there exists a positive definite hermitian form</u> φ <u>which is invariant under the</u> G-<u>action</u>, <u>and which defines the same topological structure on</u> H.

Proof. Since the mappings $s \longmapsto \pi(s)v$ are continuous, the mappings

$$s \longmapsto (\pi(s)v \mid \pi(s)w) \qquad (v,w \in H)$$

will also be continuous (continuity of the scalar product in $H \times H$). We can thus define

$$\varphi(v,w) = \int_G (\pi(s)v \mid \pi(s)w)\, ds$$

using Haar integral. It is clear that φ is hermitian and positive. Let us show that it is non-degenerate and defines the same topology on H.
Since G is compact, $\pi(G)$ is also compact in GL(H) (strong topology in this space). In particular, $\pi(G)$ is *simply bounded* whence *uniformly bounded* (Banach-Steinhaus theorem: Rudin 1973 pp.43-44). Thus there exists a positive constant $M > 0$ with

$$\|\pi(s)\, v\| \leqslant M\|v\| \qquad (s \in G,\ v \in H)\ .$$

We thus see that

$$\|v\| = \|\pi(s^{-1})\pi(s)\, v\| \leqslant M\|\pi(s)\, v\| \leqslant M^2\|v\|$$

and thus

$$M^{-1}\|v\| \leqslant \|\pi(s)\, v\| \leqslant M\|v\| \qquad .$$

Squaring and integrating over G, we find

$$M^{-2}\|v\|^2 \leqslant \varphi(v,v) \leqslant M^2\|v\|^2 \qquad .$$

Thus $\varphi(v,v) = 0$ obviously implies $\|v\| = 0$ and $v = 0$. At the same time, we see that φ and $\|\ldots\|^2$ induce equivalent topologies (equivalent norms) on H. Finally, invariance of φ comes from invariance of the Haar measure

$$\varphi(\pi(t)v,\pi(t)w) = \int_G (\pi(st)v \mid \pi(st)w)\, ds =$$

$$= \int_G f(st)\, ds = \int_G f_t(s)\, ds = \int_G f(s)\, ds = \varphi(v,w).$$

This shows that π is φ-*unitary* as desired. q.e.d.

From these propositions follows that any representation of a compact group in a Hilbert space is equivalent to a unitary one, and any *finite dimensional* representation (the dimension of a representation is of course the dimension of its representation space) is <u>completely reducible</u> (i.e. direct sum of irreducible ones).

When E is still finite dimensional, say $E = \mathbb{C}^n$, we know that any hermitian form φ can be put in diagonal form. Hence, the unitary group $U(\varphi)$ with respect to φ is conjugate to the standard unitary group $U_n(\mathbb{C})$ in $Gl_n(\mathbb{C})$. In particular, the second proposition shows that any compact subgroup G of $GL_n(\mathbb{C})$ is contained in a conjugate of $U_n(\mathbb{C})$ (apply that proposition to the identity representation $G \hookrightarrow Gl_n(\mathbb{C})$). We see thus that the conjugates $U(\varphi)$ of $U_n(\mathbb{C})$ are the *maximal compact subgroups* of $Gl_n(\mathbb{C})$. One can show similarly that the orthogonal group $O_n(\mathbb{R})$ is a maximal compact subgroup of $Gl_n(\mathbb{R})$, and all maximal compact subgroups of $Gl_n(\mathbb{R})$ are conjugate to this one.

GENERAL CONSTRUCTION OF REPRESENTATIONS

The main interest of the notion of representation of (compact) groups comes from the fact that each such group has indeed some canonical representations. We shall even construct faithful representations of any compact group G ($\pi(s) = id_E = 1_E$ only for s = e neutral element of G). For $1 \leqslant p < \infty$ let $L^p(G) = L^p(G,m)$ denote the Banach space obtained by completing the space C(G) normed with

$$\|f\|_p^p = \int_G |f(s)|^p \, ds \quad .$$

In particular $L^2(G)$ is a Hilbert space since its norm derives from the scalar product

$$(f \mid g) = \int_G \overline{f(s)} g(s) \, ds \quad .$$

In any space of functions on G, we define left translations by

$$(\ell(s)f)(x) = f(s^{-1}x)$$

(if we do not want to identify elements of $L^p(G)$ with functions, or classes of functions, we can simply extend translations from C(G) to $L^p(G)$ by continuity). Thus we have

$$\ell(s)\ell(t) = \ell(st) \qquad (s, t \in G) ,$$

and we get homomorphisms

$$\ell : G \to Gl(E) \ , \quad s \mapsto \ell(s)$$

with any $E = L^p(G)$, $1 \leqslant p < \infty$. Let us check that these homomorphisms are continuous (in our representation sense). For fixed $\varepsilon > 0$ and $f \in L^p(G)$, we can find a continuous function $h \in C(G)$ with $\|f - h\| < \varepsilon$ (the norm is the p-norm of $L^p(G)$). For s and t in G, we have

$$\|\ell(s)f - \ell(t)f\| \leq$$
$$\leq \|\ell(s)(f - h)\| + \|\ell(s)h - \ell(t)h\| + \|\ell(t)(h - f)\| .$$

But, by invariance of the Haar measure, all operators $\ell(s)$ are isometric and thus

$$\|\ell(s)(f - h)\| = \|\ell(t)(f - h)\| = \|f - h\| < \varepsilon .$$

On the other hand, since h is continuous and G is compact, h is uniformly continuous and

$$\ell(t)h \longrightarrow \ell(s)h \quad \textit{uniformly} \text{ when } t \rightarrow s \quad .$$

A fortiori, $\ell(t)h \longrightarrow \ell(s)h$ in the norm of $L^p(G)$ and we can find a neighbourhood V_s of s such that

$$t \in V_s \implies \|\ell(s)h - \ell(t)h\| < \varepsilon .$$

We shall then have $\|\ell(s)f - \ell(t)f\| < 3\varepsilon$. This proves that all mappings

$$G \longrightarrow L^p(G) , \quad s \longmapsto \ell(s)f \qquad (f \in L^p(G))$$

are continuous and the homomorphisms ℓ are representations. These are the <u>left regular representations</u> of G. The <u>right regular representations</u> of G in the Banach spaces $L^p(G)$ are defined completely similarly with

$$(r(s)f)(x) = f(xs) \qquad (f \in L^p(G)) .$$

With this definition of right translate, one has indeed $r(st) = r(s)r(t)$. One can also consider the biregular representations

$$\ell \times r \text{ of } G \times G \text{ in } L^p(G)$$

defined by

$$\{\ell \times r(s,t)(f)\}(x) = f(s^{-1}xt) \qquad (f \in L^p(G))$$

and its restriction to the diagonal $G \longrightarrow G \times G$, $s \longmapsto (s,s)$ which is the <u>adjoint representation</u> of G. It is defined by

$$\{Ad(s) f\}(x) = f(s^{-1}xs) \qquad (f \in L^p(G)) .$$

The regular representations are faithful (for any couple of distinct points s,t in G, there is a continuous function f with $f(s) = 0$ and $f(t) \neq 0$), but the adjoint representation can be trivial (this is the case when G is commutative!).

There are <u>other general examples</u> of representations. They are obtained by some *functorial constructions*. We have already alluded to the direct sum of representations (when we mentioned complete reducibility of finite dimensional representations of compact groups). Here are some examples of these constructions.

Let $\pi : G \longrightarrow Gl(E)$ and $\pi' : G' \longrightarrow Gl(E')$ be two representations. We can define the <u>external direct sum</u> representation of $G \times G'$ in $E \oplus E'$ by

$$\pi \dot{\oplus} \pi' (s,s') = \pi(s) \oplus \pi(s') \qquad (s \in G,\ s' \in G') .$$

When $G = G'$, we can restrict this external direct sum to the diagonal G of $G \times G$, obtaining the (<u>usual</u>) <u>direct sum</u> of π and π'

$$\pi \oplus \pi' : G \longrightarrow Gl(E \oplus E')$$
$$s \longmapsto \pi(s) \oplus \pi'(s) .$$

One can also define the <u>external tensor product</u> $\pi \otimes \pi'$ as a representation of $G \times G'$ in $E \times E'$ (let us assume that one of the two spaces E, E' is finite dimensional , so that this algebraic tensor product is complete : in general, some completion has to be devised). It is defined by

$$\pi \otimes \pi' (s,s') = \pi(s) \otimes \pi'(s') \qquad (s \in G,\ s' \in G') .$$

The (<u>usual</u>) <u>tensor product</u> of two representations of the same group G is again the restriction to the diagonal of the external tensor product ($G = G'$) and is simply given by

$$\pi \otimes \pi' (s,t) = \pi(s) \otimes \pi'(t) \qquad (s,\ t \in G) .$$

This is a representation of G acting in $E \otimes E'$. In particular, taking $\pi' = \pi$, one can define $\pi \otimes \pi = \pi^{\otimes 2}$ and by induction, all (<u>tensor</u>) <u>powers</u>

$$\pi^{\otimes n} = \pi \otimes \pi \otimes \ldots \otimes \pi \qquad \text{(n terms)} .$$

These powers can be restricted to symmetric (resp. anti-symmetric) tensors. We thus obtain definitions of the <u>symmetric powers</u> $S^n\pi$ (resp. <u>anti-symmetric powers</u> $\Lambda^n\pi$) of π.

When $\pi : G \longrightarrow Gl(E)$ is a given representation of G, we can also define the contragredient representation $\check{\pi}$ of π. This representation acts in the dual E' (Banach space of continuous linear forms on E) of E, and is defined by

$$\check{\pi}(s) = {}^t\pi(s^{-1}) \qquad (s \in G)$$

(since transposition reverses the order of composition of mappings ${}^t(AB) = {}^tB\,{}^tA$, it is necessary to reverse once more the operations by taking the inverse in the group : in this way $\check{\pi}(st) = \check{\pi}(s)\check{\pi}(t)$ as is required for a representation !) . When E = H is a Hilbert space, we can define the conjugate $\bar{\pi}$ of π as a representation acting in the conjugate $\bar{H}$ of H . Recall that this space $\bar{H}$ has same underlying additive group as H, but the scalar multiplication in $\bar{H}$ is twisted by complex conjugation. More precisely, the external operation of scalars in $\bar{H}$ is given by

$$(a,v) \longmapsto a\cdot v = \bar{a}v \quad \text{(we use a dot in } \bar{H}).$$

The scalar product $(.\,|\,.)^-$ of $\bar{H}$ is defined by

$$(v\,|\,w)^- = \overline{(v\,|\,w)} \quad (= (w\,|\,v)) \quad .$$

This suggests that an element v of H is written $\bar{v}$ when we consider it as element of the dual Hilbert space $\bar{H}$. With such a notation, we have

$$\overline{av} = \bar{a}\cdot\bar{v} \quad (a \in \mathbb{C}) \text{ and } (\bar{v}\,|\,\bar{w})^- = \overline{(v\,|\,w)} \quad .$$

The identity map $H \longrightarrow \bar{H}$, $v \longmapsto \bar{v}$ is an *anti-isomorphism*. The conjugate of π is simply defined by $\bar{\pi}(s) = \pi(s)$ *in* $\bar{H}$. Since the (complex vector) subspaces of H and $\bar{H}$ are *the same* by definition, π and $\bar{\pi}$ are reducible or irreducible simultaneously. However, it is important to distinguish these two representations (in particular, they are not always *equivalent*). Any orthonormal basis (e_i) of H is also an orthonormal basis of $\bar{H}$, but a decomposition $v = \sum v_i e_i$ in H gives rise to the decomposition

$$\bar{v} = \sum \bar{v}_i \cdot \bar{e}_i \quad \text{(complex conjugate components) in } \bar{H} \ .$$

Thus the *matrix representations* associated with π and $\bar{\pi}$ in the bases (e_i) $(= (\bar{e}_i))$ of H (resp. $\bar{H}$) are complex conjugate to one another.

When π is unitary, the contragredient $\check{\pi}$ and the conjugate $\bar{\pi}$ of π are equivalent (ex.2 below).

There is another general procedure for constructing representations of G : it is *induction starting from a representation of a (closed) subgroup* of G. We shall give its definition in sec.8 below .

EXERCISES

1. *Show that the left and right representations* ℓ *and* r *of a group* G *(in any* $L^p(G)$ *space) are equivalent.*
(Hint : consider the mapping $f \longmapsto \check{f}$ *defined on any space of functions on* G *where* $\check{f}(x) = f(x^{-1})$*.)*

2. *Let* π *be a unitary representation of a group* G*. Show that* $\check{\pi}$ *and* $\bar{\pi}$ *are equivalent.*
(Hint : Riesz' theorem asserts that $\bar{H} \longrightarrow H'$, $\bar{v} \longmapsto (v \mid .)$ *is a complex isomorphism.)*

3. *If* π *and* π' *are two representations of the same group* G *(acting in respective Hilbert spaces* H *and* H'*), show that the matrix coefficients of* $\pi \otimes \pi'$ *(with respect to bases* (e_i) *of* H*,* (e'_j) *of* H' *and* $(e_i \otimes e'_j)$ *of* $H \otimes H'$ *) are products of matrix coefficients of* π *and* π' *(Kronecker product of matrices).*

4. *Let* 1_n *denote the identity representation of a group* G *in dimension* n *(the space of this identity representation is thus* $\mathbb{C}^n$ *and* $1_n(s) = id_{\mathbb{C}^n}$ *for every* $s \in G$ *). Show that for any representation* π *of* G*,*

$$\pi \otimes 1_n \text{ is equivalent to } \pi \oplus \pi \oplus \ldots \oplus \pi \text{ (n terms) .}$$

5. *(Schur's lemma) Let* k *be an algebraically closed field,* V *a finite dimensional vector space over* k *and* Φ *any irreducible family of operators in* V *(the only invariant subspaces, relatively to all operators belonging to* Φ*, are* $\{0\}$ *and* V *) . Then, if an operator* A *commutes with all operators of* Φ *,* A *is a multiple of the identity operator (i.e.* A *is a scalar operator). Hint: take an eigenvalue* a *(in the algebraically closed field* k*) of* A *and consider* $A - aI$*, which still commutes with all operators of* Φ*. Show that* $\mathrm{Ker}(A - aI)$ $(\neq \{0\})$ *is an invariant subspace.*

3 A GEOMETRICAL APPLICATION

In this section, we shall answer the following question : Which are the bounded symmetrical star-shaped (with respect to the origin) bodies in $\mathbb{R}^3$ having all plane sections through the origin of equal area ?

Obviously spheres (with center at the origin) have the required property. We shall show that there are no others...

Let K be any bounded star-shaped (with respect to the origin) subset of $\mathbb{R}^3$. We shall assume that K is closed (hence compact) and describe this body by the function on the unit sphere

$$f(\vec{r}) = \operatorname{Sup}\{\lambda > 0 : \lambda\vec{r} \in K\} \qquad (\|\vec{r}\| = 1).$$

Since we want to evaluate *areas*, we assume that f is *measurable* hence *integrable* (it is true that this follows from the fact that K is closed, but our object is not to insist on these measure theoretic questions...).

Areas can be evaluated in polar coordinates according to the well known formula

$$\tfrac{1}{2}\int_0^{2\pi} \rho(\varphi)^2\, d\varphi \quad .$$

In our case, if $\|\vec{r}\| = 1$, i.e. $\vec{r}$ on the unit sphere $S^2 \subset \mathbb{R}^3$, we denote by $Af(\vec{r})$ the area of the planar section (of K) orthogonal to $\vec{r}$. Using the circle $\gamma_{\vec{r}} = \{\vec{s} : \vec{s} \perp \vec{r}, \|\vec{s}\| = 1\} \subset S^2$, we can express this area by

$$Af(\vec{r}) = \tfrac{1}{2}\oint_{\vec{s}\in\gamma_{\vec{r}}} f(\vec{s})^2\, ds \qquad (ds = \|d\vec{s}\|).$$

Thus our assumption on K means that Af is a constant function (and we must show that f is constant when it is even,i.e. K is symmetrical).

Let us study more systematically the operator

$$J : f \longmapsto Jf \quad , \quad Jf(\vec{r}) = \oint_{\gamma_{\vec{r}}} f(\vec{s})\, ds \quad .$$

We shall show that J is *one to one on even functions* (in other words, the kernel of J consists precisely of odd functions). As a consequence, the only functions f such that Jf = constant, are the constant functions f (describing spheres).

Even more precisely, we shall determine all eigenvalues of the operator J in $L^2(S^2)$, or in

$$L^2_{even}(S^2) = L^2(S^2/\pm 1) = L^2(\mathbb{P}^2) \quad .$$

Thus $\mathbb{P}^2$ represents the real projective plane. Technically, we need a first lemma.

(3.1) Lemma. Assume that the operator J is given on continuous functions f by the above integral formula. Then J is continuous in quadratic mean and extends continuously to the space $L^2(\mathbb{P}^2)$. Moreover, if $f \in L^2(\mathbb{P}^2)$, the integral expression of J is still valid *for nearly all* $\vec{r}$.

Proof. Let $\vec{e}_1$ be the first element of the canonical basis of $\mathbb{R}^3$ and γ_1 denote the orthogonal circle to $\vec{e}_1$ (on S^2). Take an arbitrary rotation $g \in SO_3(\mathbb{R})$ and put $g\,\vec{e}_1 = \vec{r}$, so that $\gamma_{\vec{r}} = g\,\gamma_1$. We can make the isometric change of variable (transformation with Jacobian 1) $g\,\vec{s}_1 = \vec{s}$ in the integral for $Jf(\vec{r})$

$$Jf(\vec{r}) = \oint_{\gamma_{\vec{r}}} f(\vec{s})\,ds = \oint_{\gamma_1} f(g\,\vec{s}_1)\,ds_1 \qquad \text{(f continuous)} .$$

Thus

$$|Jf(\vec{r})| \leqslant \oint_{\gamma_1} |f(g\,\vec{s}_1)|\,ds_1$$

and using the Cauchy-Schwarz inequality in $L^2(\gamma_1)$ for the scalar product of the two functions

$$\text{constant } 1 \quad \text{and} \quad \vec{s}_1 \longmapsto f(g\,\vec{s}_1) \quad ,$$

we conclude

$$|Jf(g\vec{e}_1)|^2 = |Jf(\vec{r})|^2 \leqslant \|1\|^2_{\gamma_1} \oint_{\gamma_1} |f(g\,\vec{s}_1)|^2\,ds_1 \quad .$$

Integrating moreover over $g \in SO_3(\mathbb{R})$, using the invariant measure coming from the spherical measure (we only use the formula

$$\int_G F(g\,\vec{r}_1)\,dg = \frac{1}{4\pi}\iint_{S^2} F(\vec{r})\,d\sigma \qquad (G = SO_3(\mathbb{R}))$$

for the normalized measure of the rotation group... the Haar measure of this rotation group G will be derived in sec. 9 below) we obtain

$$\|Jf\|^2_{S^2} = 4\pi \int_G |Jf(g\vec{e}_1)|^2\,dg \leqslant 4\pi\, 2\pi \oint_{\gamma_1} ds_1 \int_G dg\,|f(g\,\vec{s}_1)|^2$$

(we have used $\|1\|^2_{\gamma_1} = 2\pi$ and Fubini's theorem for an integrable function!).

As the value

$$4\pi \int_G dg |f(g\vec{s}_1)|^2 = \|f\|^2$$

is independent of $\vec{s}_1$, we conclude

$$\|Jf\|^2 \leqslant (2\pi)^2 \|f\|^2 \quad .$$

Hence J can indeed be continuously extended from $C(S^2)$ to $L^2(S^2)$ with $\|J\| \leqslant 2\pi$ (in fact, $\|J\| = 2\pi$ since $J(1) = 2\pi$). q.e.d.

To find invariant subspaces of J in $L^2(S^2)$ or in $L^2(\mathbb{P}^2)$, we observe that J commutes to rotations (hence is a G-*morphism*). We define indeed rotations of functions by the usual formula for left translations, namely

$$\ell(g)(f) = {}_g f \quad , \quad {}_g f(g\vec{r}) = f(\vec{r}) \text{ or } {}_g f(\vec{r}) = f(g^{-1}\vec{r}) \ .$$

Then

$$Jf(\vec{r}) = \int_{\gamma_{\vec{r}}} f(\vec{s})\, ds = \int_{\gamma_{\vec{r}}} {}_g f(g\vec{s})\, ds$$

is computed using the isometric change of variable $\vec{t} = g\vec{s} \in g\gamma_{\vec{r}} = \gamma_{g\vec{r}}$

$$Jf(\vec{r}) = \int_{\gamma_{g\vec{r}}} {}_g f(\vec{t})\, dt = J({}_g f)\,(g\vec{r}) \quad .$$

But the left term is

$$Jf(\vec{r}) = {}_g(Jf)(g\vec{r}) \quad ,$$

so that

$$J({}_g f) = {}_g(Jf) \quad , \quad J\cdot\ell(g) = \ell(g)\cdot J \qquad (g \in G = SO_3(\mathbb{R})).$$

This is the intertwining property and it is certainly appropriate to decompose the left regular representation of G in $L^2(\mathbb{P}^2)$. For an integer $n \in \mathbb{N}$, define the subspace

$$M_n = \{ p|_{S^2} : p \text{ is a homogeneous polynomial of degree } n \}$$

of the space of continuous functions over the sphere S^2. Since the restriction of the homogeneous polynomial $x^2 + y^2 + z^2$ of degree 2 on S^2 is identically 1, we see that $M_n \subset M_{n+2}$ $(n \in \mathbb{N})$. It is also clear that *even* n's give spaces of even functions and thus subspaces of $L^2(\mathbb{P}^2)$. Let us define

$$V_n = \text{orthogonal supplement of } M_{n-2} \text{ in } M_n \ (\subset L^2(\mathbb{P}^2))\ .$$

We shall then have (for n even)

$$M_n = V_n \oplus M_{n-2} = V_n \oplus V_{n-2} \oplus M_{n-4} =$$
$$= V_n \oplus V_{n-2} \oplus \dots \oplus V_2 \oplus V_o \qquad (V_o = \mathbb{C}).$$

It is also clear that

$$\bigcup_{n \text{ even} \geqslant 0} M_n = \bigoplus_{n \text{ even} \geqslant 0} V_n \qquad \text{(orthogonal sum)}$$

is a *subalgebra* of the algebra of continuous functions on $\mathbb{P}^2$ (algebra of continuous, even functions on S^2). This subalgebra contains the constants, and is stable under complex conjugation. Moreover, since the restrictions to S^2 of

$$x^2 , y^2 , z^2 , xy , yz , zx \qquad (\text{in } M_2)$$

already separate points of the compact space $\mathbb{P}^2 = S^2/\pm 1$, the subalgebra of even polynomial functions is dense in $C(\mathbb{P}^2)$ (Stone-Weierstrass theorem). In particular,

$$\bigoplus_{n \text{ even} \geqslant 0} V_n \text{ is dense in } L^2(\mathbb{P}^2)$$

and the completion of this orthogonal direct sum (Hilbert sum) gives a decomposition of $L^2(\mathbb{P}^2)$:

$$L^2_{\text{even}}(S^2) = L^2(\mathbb{P}^2) = \widehat{\bigoplus_{n \text{ even} \geqslant 0}} V_n .$$

Each M_n is obviously finite dimensional and G-invariant (with respect to the left regular representation) : the same must hold for V_n (cf. (2.1)).

(3.2) <u>Lemma</u>. <u>The dimension of</u> V_n <u>is</u> $2n + 1$ $(n \in \mathbb{N})$.

<u>Proof</u>. The homogeneous monomials

$$\begin{cases} x^n, \\ x^{n-1}y, \; x^{n-1}z, \\ x^{n-2}y^2, \; x^{n-2}yz, \; x^{n-2}z^2, \\ \dots \\ y^n, \; y^{n-1}z, \; \dots , \; z^n \end{cases}$$

constitute a basis of M_n . The number of elements in this basis is

$$1 + 2 + \dots + (n+1) = \tfrac{1}{2}(n+1)(n+2) .$$

Consequently,

$$\dim V_n = \dim M_n - \dim M_{n-2} = \tfrac{1}{2}(n+1)(n+2) - \tfrac{1}{2}n(n-1) = 2n + 1 \quad \text{q.e.d.}$$

We intend to prove that the V_n's (even n's) are irreducible G-modules (i.e. the left regular representation is irreducible in each such subspace of $L^2(\mathbb{P}^2)$) . To see this, let us consider the unit sphere S^2 as G (= $SO_3(\mathbb{R})$)-orbit of $\vec{e}_1$ (for the canonical action in $\mathbb{R}^3$). The stabilizer of this vector $\vec{e}_1$ for this action is the closed (hence compact) subgroup $K = SO_2(\mathbb{R})$ (operating in the orthogonal plane to $\vec{e}_1$). We make identifications

$$\begin{array}{ccc} G \longrightarrow S^2 & g \longmapsto g\vec{e}_1 & L^2(S^2) \\ \searrow \;\; /\!\!/\!\!\wr & \searrow \;\; \nearrow & \wr\!\! \| \\ G/K & gK & L^2(G/K) \end{array}$$

(3.3) Lemma. <u>Any non-zero</u> G-<u>invariant subspace</u> V <u>of</u> $L^2(G/K)$ <u>has some non-zero</u> K-<u>fixed vectors</u> :

$$V^K = \{ f : \ell(k)f = f \quad (k \in K)\} \neq \{0\} .$$

<u>Proof</u>. The K-fixed vectors $f \in L^2(G/K)$ correspond to functions over the unit sphere S^2 which are invariant under rotations around the $\vec{e}_1$-axis. These functions depend only on x (remember $y^2 + z^2 = 1 - x^2$ on S^2). Take any $0 \neq f \in V$ and replace (if necessary) f by a suitable ${}_gf$ in order to have $f(\vec{e}_1) \neq 0$. Define $f^K \in V^K$ by averaging along circles parallel to γ_1 :

$$f^K(\vec{r}) = \int_K f(k\vec{r})\,dk \qquad \left(\int_K dk = 1 \right) .$$

As all left translates ${}_kf$ $(k \in K)$ are still in the G-invariant subspace V, we see that $f^K \in V$. Moreover, $f^K(\vec{e}_1) = f(\vec{e}_1) \neq 0$ implies $f^K \neq 0$. One checks without difficulty (using invariance of the Haar measure of K) that $0 \neq f^K \in V^K$. q.e.d.

Observe that the functions f on S^2 which are invariant under K-rotations are identified with functions over $K\backslash G/K$. Thus

$$V^K = V \cap L^2(K\backslash G/K) .$$

(3.4) <u>Lemma</u>. <u>The</u> V_n (n <u>even</u> ≥ 0) <u>are irreducible subspaces of</u> $L^2(\mathbb{P}^2)$.

<u>Proof</u>. In any direct sum decomposition

$$V_n = V'_n \oplus V''_n \qquad (V'_n \text{ and } V''_n \text{ G-invariant subspaces of } V_n),$$

we must have

$$V_n^K = V'^K_n \oplus V''^K_n .$$

If we can show that $\dim V_n^K = 1$, we shall necessarily have $V'^K_n = 0$ or $V''^K_n = 0$. By the preceding lemma (3.3) it would follow $V'_n = 0$ or $V''_n = 0$ whence the irreducibility. Thus we shall prove that V_n^K has dimension 1. But elements of this space V_n^K are functions which depend only on x. In M_n, such polynomial functions must be linear combinations of

$$x^n,\ x^{n-2}(y^2+z^2) = x^{n-2}(1-x^2),\ \ldots \quad .$$

A basis for M_n^K is thus constituted of polynomials

$$x^n,\ x^{n-2},\ \ldots,\ x^2,\ 1 \quad .$$

Hence $\dim M_n^K = \frac{1}{2}n + 1$ (recall that n is even !) and also

$$\dim V_n^K = \dim M_n^K - \dim M_{n-2}^K = 1 \quad .$$

This completes the proof, but we observe that the formula $\dim M_n^K = \frac{1}{2}n + 1$ corresponds to a decomposition

$$M_n^K = V_n^K \oplus V_{n-2}^K \oplus \ldots \oplus V_2^K \oplus V_o^K \qquad (V_o^K = V_o = \mathbb{C})$$

where each V_i^K is a *line* (i.e. has dimension 1). q.e.d.

From the analysis just made, it follows that the intertwining operator (or G-morphism) J must leave each V_n subspace (n even $\geqslant 0$) *invariant* and operate like a *scalar* in each of them (Schur's lemma, cf. ex. 5 of sec.2). We have to determine these constants (to show that none of them is 0 !)

$$J_n = J\big|_{V_n} = \lambda_n \,\mathrm{id}_{V_n} \qquad (n \text{ even } \geqslant 0).$$

(3.5) Lemma. The scalar product of two functions f_1, f_2 which depend only on x on the unit sphere S^2 (i.e. $f_i \in L^2(K\backslash G/K)$) is given by

$$(f_1 \mid f_2) = \iint_{S^2} \bar{f}_1 f_2 \, d\sigma =$$

$$= 2\pi\int_0^{\pi} \overline{f_1(\cos\theta)}\, f_2(\cos\theta)\sin\theta \, d\theta = 2\pi\int_{-1}^{1} \overline{f_1(x)}\, f_2(x)\, dx \quad .$$

The proof of this lemma is obvious (exercise !). From this lemma follows that we obtain a sequence of orthogonal polynomials $P_n \in V_n^K$ (n even $\geqslant 0$) by orthogonalization of the sequence of even powers $1, x^2, \ldots$ for the usual scalar product on $[-1,+1]$ ("isomorphic" to $K\backslash G/K$ up to the $\pm$ sign). We know that we can take the *Legendre polynomials* (of even index). These polynomials are given by the *Rodrigue's formula* (cf. Robert 1973)

$$P_n(x) = \frac{1}{2^n n!} \frac{d^n}{dx^n}(x^2 - 1)^n \qquad (n \in \mathbb{N}).$$

Successive integrations by parts show indeed that these n^{th} derivatives define a sequence of polynomials with

$$\deg(P_n) = n \quad \text{and} \quad P_n \text{ orthogonal to } x^m \text{ for } m < n \quad .$$

On the other hand, Leibniz' formula for the nth derivative of a product can be used with

$$(x^2 - 1)^n = (x+1)^n (x-1)^n$$

and we thus see easily that $P_n(1) = 1$ (this is the usual *normalization* condition, justifying the introduction of the factor $(2^n n!)^{-1}$ in front of the nth derivative). Thus we have

$$V_n^K = \mathbb{C}\, P_n \qquad (n \text{ even} \geqslant 0)$$

and the computation of the eigenvalue λ_n of the operator $J_n = J|_{V_n}$ is reduced to finding the value of $J(P_n)$ at *one particular point*:

$$\lambda_n = \lambda_n P_n(1) = J(P_n)(\vec{e}_1) \quad .$$

But by definition of J, we have

$$J P_n(\vec{e}_1) = \oint_{\gamma_1} P_n(\vec{s}_1)\, ds_1 = \int_0^{2\pi} P_n(0)\, d\varphi = 2\pi\, P_n(0)$$

(we identify P_n to a function on S^2 in the obvious way...) . It only remains to determine the value of P_n at the origin (this value is 0 for odd n's since P_n has the same parity as n, and only even n's are interesting for us). This value $P_n(0)$ is the constant term of the polynomial P_n, in the nth derivative; this term comes from the nth power of x in $(x^2 - 1)^n$. The binomial formula gives this interesting coefficient of x^n : it is $(-1)^{\frac{1}{2}n} \binom{n}{n/2}$. Summing up, we find that $P_n(0)$ (n even $\geqslant 0$) is the product of

$(2^n n!)^{-1}$ (normalization coefficient in Rodrigue's formula),

$n!$ (coming from the nth derivative of x^n)

and

$(-1)^{n/2} \binom{n}{n/2}$ (binomial coefficient of x^n in $(x^2 - 1)^n$) .

Finally,

$$\lambda_n = (-1)^{n/2}\, 2\pi \frac{(n-1)(n-3) \ldots 3 \cdot 1}{n(n-2) \ldots 4 \cdot 2} \neq 0 \qquad (n \text{ even} \geqslant 0) \quad .$$

EXERCISE

Generalize the results of this section for the canonical action of $G = SO_n(\mathbb{R})$ *on the unit sphere* $S^{n-1} \subset \mathbb{R}^n$ *as follows.*

1) Denote by $M_\ell(n)$ *the space of homogeneous polynomials of degree* ℓ *on* $\mathbb{R}^n$. *Prove*

$$\dim M_\ell(n) = \frac{(\ell+n-1)!}{\ell!(n-1)!} .$$

(Here is G. B. Folland's explanation for this. The number of monomials of degree ℓ *is the number of sequences* $(i_1,\dots,i_n)$ *of non-negative integers with* $i_1+\dots+i_n = \ell$. *Line up* ℓ *black balls in a row and divide them into n groups of consecutive balls with cardinalities* $i_1, \dots, i_n$. *To mark the division between two adjacent groups, interpose a white ball between two black balls : for this purpose, we need n-1 white balls. The number of ways we can make such a configuration is the number of ways of taking* $\ell+n-1$ *black balls and choosing n-1 of them to be painted white !)*

2) By restriction to the unit sphere, we write

$$M_\ell(n) = M_{\ell-2}(n) + V_\ell(n) \hookrightarrow L^2(S^{n-1})$$

with an orthogonal complement $V_\ell(n)$ *of dimension*

$$\dim V(n) = (2\ell + n - 2) \frac{(n+\ell-3)!}{\ell!(n-2)!} .$$

(One can show that this space $V_\ell(n)$ *consists of restrictions of the harmonic polynomials in* $M_\ell(n)$.*)*

3) Prove that the natural representation of $SO_n(\mathbb{R})$ *in* $V_\ell(n)$ *is irreducible. (Embed* $K = SO_{n-1}(\mathbb{R})$ *in* $G = SO_n(\mathbb{R})$ *by means of the last n-1 coordinates and prove that the space of K-fixed vectors in* $V_\ell(n)$ *has dimension 1.) Conclude that the decomposition of* $L^2(S^{n-1})$ *into irreducible components is given by the Hilbert sum* $\bigoplus_{\ell \geq 0} V_\ell(n)$.

4) Identifying K-invariant functions with functions of $x = x_1 = \cos\theta$ *with* θ = *angle between* $\vec{r}$ *and* $\vec{e}_1$, *show that* $V_\ell(n)^K$ *consists of multiples of the Gegenbauer (or ultraspherical) polynomials*

$$C_\ell^{\frac{1}{2}n-1}$$

(Recall that the system of polynomials (C_ℓ^ν) *is orthogonal with respect to the density* $(1-x^2)^{\nu-\frac{1}{2}}$ *on the interval* $[-1,1]$ *.)*

4 FINITE-DIMENSIONAL REPRESENTATIONS OF COMPACT GROUPS (PETER-WEYL THEOREM)

In sec. 2, we have shown that all compact groups have *faithful* representations. For that purpose, the regular representations were constructed and examined. However, these last representations are *infinite dimensional* in general (i.e. when G is infinite). In particular, this does not prove the existence of irreducible representations (different from the identity in dimension 1) for compact groups.

This section is devoted mainly to the proof of the following basic result (and its consequences).

Theorem (Peter-Weyl). Let G be a compact group. For any $s \neq e$ in G, there exists a finite dimensional, irreducible representation π of G such that $\pi(s) \neq \mathrm{id}$. .

Since certain compact groups have no faithful finite dimensional representations (groups with arbitrarily small subgroups are in this class when infinite), this result is the best possible. This theorem is sometimes stated in the following terms : all compact groups have *enough* finite dimensional representations, or: all compact groups have a *complete system* of (irreducible) finite dimensional representations. As we have already seen that all finite dimensional representations of compact groups are completely reducible, the theorem will already be proved if we show that for $s \neq e$ in G, there exists a finite dimensional representation π with $\pi(s) \neq \mathrm{id}$. .

The proof of the preceding theorem is based on the spectral properties of compact hermitian operators in Hilbert spaces. Let us review the main points needed.

Lemma 1. Let G be a compact group , $k : G \times G \longrightarrow \mathbb{C}$ be a continuous function and $K : L^2(G) \longrightarrow C(G)$ be the operator with *kernel* k :

$$(Kf)(x) = \int_G k(x,y)\, f(y)\, dy \quad .$$

Then K is a *compact* operator, and if moreover $k(y,x) = \overline{k(x,y)}$ (identically on $G \times G$) K is hermitian as operator $L^2(G) \longrightarrow C(G) \subset L^2(G)$.

Let us quickly recall the proof of this lemma. Denote by k_x the functions $y \mapsto k(x,y)$ on G. Thus we have

$$(Kf)(x) = \int_G k_x(y)\, f(y)\, dy = (\overline{k_x} \mid f) \quad .$$

Since $x \mapsto k_x$ is a continuous mapping $G \longrightarrow C(G) \hookrightarrow L^2(G)$ (k is *uniformly continuous* on the compact space $G \times G$), we deduce that all functions Kf ($f \in L^2(G)$) are continuous : $Kf \in C(G)$. Moreover, the Cauchy-Schwarz inequality gives

$$|Kf(x)| \leqslant \|k_x\| \|f\| \leqslant \operatorname*{Max}_{G\times G} |k| \cdot \|f\| \quad .$$

In particular, when f remains in the unit ball $\|f\| \leqslant 1$ of $L^2(G)$ and x is fixed, the numbers Kf(x) remain in a bounded (i.e. relatively compact) subset of $\mathbb{C}$. On the other hand, the set Φ of all Kf for $\|f\| \leqslant 1$ is equicontinuous :

$$|Kf(s) - Kf(t)| \leqslant \|k_s - k_t\| \cdot \|f\| \leqslant \|k_s - k_t\|$$

is arbitrarily small for s sufficiently close to t (independently of f in the unit ball of $L^2(G)$). Ascoli's theorem (cf. sec.1) shows that the set of Kf for $\|f\| \leqslant 1$, i.e. the image of the unit ball of $L^2(G)$ by the operator K, is relatively compact in C(G) (with the uniform norm in this last space). A fortiori, the operator K is a compact operator of $L^2(G)$ into itself. Finally, a formal computation shows that the adjoint K* of K is also an integral operator, with kernel $k^*(x,y) = \overline{k(y,x)}$.

Lemma 2. Let K be a compact operator (in some Hilbert space H). Then the spectrum S of K consists of eigenvalues and possibly of 0. Each eigenspace H_λ with respect to a non-zero eigenvalue $\lambda \in S$ is finite-dimensional. When moreover K is hermitian, $S \subset \mathbb{R}$,

$$\|K\| = \operatorname*{Sup}_{\lambda \in S} |\lambda|$$

and eigenspaces associated to distinct eigenvalues are orthogonal

$$H_\lambda \perp H_\mu \quad \text{for} \quad \lambda \neq \mu \quad \text{in } S - \{0\} \quad .$$

We shall not re-prove this classical theorem which can be found in any functional analysis book (cf. Rudin 1973 pp. 97, or Dieudonné 1960. Chap.XI where the more general case of Banach spaces is considered).

The finiteness of the dimension of the H_λ (for $\lambda \neq 0$) is however so easily shown that I recall the argument. Let B denote the (closed) unit ball of H and $B_\lambda = B \cap H_\lambda$ the (closed) unit ball of H_λ. Then, since K is compact, $K(B_\lambda)$ must be relatively compact in H_λ. But this set

$$K(B_\lambda) = \lambda B \cap H_\lambda$$

is the (closed) ball of radius $|\lambda|$ in H_λ. In particular, H_λ must be a locally compact space : it is finite dimensional.

We shall use lemma 2 above in the following form :

If $K \neq 0$ is compact and hermitian, then K has a non-zero eigenvalue λ and the corresponding eigenspace H_λ is finite-dimensional.

We can now prove the main theorem (stated at the beginning of this sec.). Assume that $s \neq e$ in G and take an open symmetric neighbourhood $V = V^{-1}$ of e in G such that $s \notin V^2$. There exists a positive continuous function f on G such that

$$f(e) > 0 \ , \quad f(x) = f(x^{-1}) = \check{f}(x) \ , \quad \mathrm{Supp}(f) \subset V$$

(Supp(f) denotes the *support* of f : it is the complement of the largest open set in which f vanishes). Replacing if necessary f by $f + \check{f}$, we can assume that $f = \check{f}$. Let us examine the function $\varphi = f * f$ defined by

$$\varphi(x) = \int_G f(y)\, f(y^{-1}x)\, dy \qquad .$$

Obviously, the support of this new function φ is contained in V^2 and

$$\varphi(s) = 0 \ (s \notin V^2) \ , \quad \varphi(e) = \|f\|^2 > 0 \ .$$

A fortiori we see that $\ell(s)\varphi \neq \varphi$. But the operator K with kernel $k(x,y) = f(y^{-1}x)$ is compact (lemma 1) and the convergence of

$$f = f_o + \sum f_i \qquad , \ f_i \in \mathrm{Ker}(K - \lambda_i I) = H_i \ (\lambda_i \in \mathrm{Spec}(K))$$

(in quadratic mean) implies that

$$\varphi = Kf = \sum Kf_i = \sum \lambda_i f_i \quad \left(f_i = \frac{1}{\lambda_i} Kf_i \in \mathrm{Im}(K) \subset C(G) \right).$$

(with a uniform convergence). Since $\ell(s)\varphi \neq \varphi$, we must have $\ell(s)f_i \neq f_i$ for one index i (at least). But the definition of the kernel k shows that

$$k(sx,sy) = k(x,y) = f(y^{-1}x) \qquad\qquad (s,x \text{ and } y \text{ in } G).$$

The consequence of these identities is the translation invariance of all the eigenspaces H_i of K. The left regular representation restricted to a suitable finite dimensional subspace H_i (any i with $\ell(s)f_i \neq f_i$ will do) will furnish an example of a finite dimensional representation π with $\pi(s) \neq e$. q.e.d.

The corollaries of the main theorem are numerous and important.

(4.1) Corollary. A compact group is commutative if and only if all its finite dimensional irreducible representations have dimension 1.

Proof. Since $Gl_1(\mathbb{C}) = \mathbb{C}^\times$ is commutative, any dimension 1 representation is trivial on the commutator subgroup $[G,G]$ of G. If all finite dimensional irreducible representations of G have dimension 1, the theorem implies that all commutators must coïncide with the neutral element of G and thus, G must be commutative. Conversely, assume that the (compact) group G is commutative, and take a finite dimensional irreducible representation π of G. Since any $\pi(s)$ commutes to all $\pi(t)$ $(t \in G)$, Schur's lemma (Ex.5 of sec.2) implies that $\pi(s)$ is a scalar operator (for all $s \in G$). Thus the whole action of G on $H = H_\pi$ is by dilatation: all subspaces of H are G-invariant. The irreducibility of π requires that H has dimension 1. q.e.d.

(4.2) Corollary (Peter-Weyl). Any continuous function on a compact group is a uniform limit of (finite) linear combinations of coefficients of irreducible representations.

Proof. Let π be a (finite dimensional) irreducible representation of the compact group G and take a basis in the representation space of π in order to be able to identify $\pi: G \longrightarrow Gl_n(\mathbb{C})$, the coefficients of π being the continuous functions

$$c_j^i \; : \; g \mapsto c_j^i(g) = (e_i \mid \pi(g)e_j)$$

on G. In fact, more generally, if u and $v \in H$ we can define the (function) coefficient c_v^u of π on G by

$$g \mapsto c_v^u(g) = (u \mid \pi(g)v)$$

(these functions are obviously finite linear combinations of the previously defined matrix coefficients c_j^i). Introduce the subspace $V(\pi)$ of $C(G)$ spanned by the c_j^i, or equivalently by all c_v^u $(u,v \in H_\pi)$.

Observe that the subspaces of C(G) attached in this way to two *equivalent* representations π and π' *coïncide*: $V(\pi) = V(\pi')$. Thus we can form the algebraic sum (a priori this algebraic sum is not a direct sum)

$$A_G = \bigoplus V_\pi \subset C(G)$$

where the summation index π runs over all (classes of) finite dimensional irreducible representations of G. The corollary can be restated in the following form

A_G *is a dense subspace of the Banach space* C(G) *(uniform norm)*.

But this algebraic sum A_G is a *subalgebra* of C(G) (the product of two continuous functions being the usual pointwise product) : the product of the coefficients

$$c_v^u \text{ of } \pi \quad \text{and} \quad \gamma_t^s \text{ of } \sigma$$

is a coefficient of the representation $\pi \otimes \sigma$ (the coefficient of this representation with respect to the two vectors $u \otimes s$ and $v \otimes t$). Taking π and σ to be finite dimensional irreducible representations of G , $\pi \otimes \sigma$ will be finite dimensional, hence completely reducible and all its coefficients (in particular the product of c_v^u and γ_t^s) are finite linear combinations of coefficients of (finite dimensional) irreducible representations of G. This subalgebra A_G of C(G) contains the constants, is stable under complex conjugation (because π is irreducible precisely when $\bar{\pi}$ is irreducible) and *separates points of* G by the main theorem. The Stone-Weierstrass theorem furnishes the conclusion. q.e.d.

Observe that on a compact group the Stone-Weierstrass theorem can be proved by convolution (regularization). Thus, the preceding proof can be made in a more elementary way. We give this alternative at the end of the section (after corollary (4.5)).

(4.3) Corollary. For each neighbourhood V of the neutral element in G, there exists a finite dimensional representation with kernel contained in V (this kernel is a closed invariant subgroup of G).

Proof. We can assume that V is open, hence G - V is compact. For each $x \in G - V$, let us construct a representation π_x as in the main theorem : $\pi_x(x) \neq \text{id}$. . By continuity of π_x , we shall still have $\pi_x(y) \neq \text{id}$.

for all y in a neighbourhood V_x of x. The interiors of the V_x's make up an open covering of the compact space $G - V$. We can select a finite sub-covering $(V_i) = (V_{x_i})$ corresponding to a finite number of points x_i. Let us denote by π_i the representation corresponding to x_i

$$\pi_i(y) \neq \text{id. for all } y \in V_i \ .$$

The direct sum $\pi = \bigoplus_i \pi_i$ of the π_i's will be such that $\pi(y) \neq$ id. for all $y \in \bigcup V_i$. Hence the kernel of π is contained in the neighbourhood V of e. q.e.d.

The next corollary is due to J. von Neumann. It gives an answer to Hilbert's fifth problem in the case of compact groups (the general case has obtained a definitive answer by the joint work of Montgomery and Zippin).

(4.4) Corollary. Let G be a compact group. The following conditions are equivalent.

i) There is a neighbourhood V of the neutral element e of G containing no closed invariant subgroup different from $\{e\}$.

ii) There is a faithful finite dimensional representation of G and G is isomorphic to a closed subgroup of a unitary group $U_n(\mathbb{C})$.

iii) G is a *real Lie group* (with a finite number of connected components).

Proof. The implication i) $\Longrightarrow$ ii) follows immediately from (4.3) above. To see that ii) $\Longrightarrow$ iii), it is enough to remember that a closed subgroup of a Lie group is a Lie group, and to apply this result to the real Lie group $U_n(\mathbb{C})$ (also observe that since G is compact, any continuous injective map $G \longrightarrow U_n(\mathbb{C})$ is a homeomorphism into). In this case, one could even see that G is an *algebraic group*. Finally, the implication iii) $\Longrightarrow$ i) is known classically. In our case, we can use the exponential map

$$M_n(\mathbb{R}) \longrightarrow Gl_n(\mathbb{R}) \ , \quad A \longmapsto \exp(A) = \sum_{n \geq 0} A^n/n! \ .$$

This map is a local diffeomorphism in the neighbourhood of $0 \in M_n(\mathbb{R})$: there exists a neighbourhood V of the zero matrix of $M_n(\mathbb{R})$ for which $\exp : V \xrightarrow{\sim} \exp(V)$ (this is a neighbourhood of 1_n in $Gl_n(\mathbb{R})$). As $\exp(nA) = (\exp A)^n$, a subgroup containing a non-trivial element $s \in \exp(V)$

contains all its powers, and these powers cannot stay in a prescribed neighbourhood $W \subset \exp(V)$ of 1_n in $Gl_n(\mathbb{R})$. q.e.d.

(4.5) Corollary. Any compact group is an inverse limit of a family of compact Lie groups (hence a closed subgroup of a product of compact Lie groups).

Proof. Let us take a fundamental system (V_i) of neighbourhoods of the neutral element in G (we can take a countable such system if G is metrizable). For each i, choose a finite dimensional representation π_i of G with kernel contained in V_i. Since inclusions $V_i \subset V_j$ do not necessarily lead to inclusions $\mathrm{Ker}(\pi_i) \subset \mathrm{Ker}(\pi_j)$ (and since the index set can be very large...) we proceed as follows. Take as new index set J = = set of finite parts of the preceding index set I, and for any $\iota \in J$, define

$$\pi_\iota = \bigoplus_{i \in \iota} \pi_i \quad , \quad V_\iota = \bigcap_{i \in \iota} V_i \quad .$$

With respect to *inclusion* (*in* I), the family $(\pi_\iota)_J$ is an inverse system of representations of G : for ι and ι' in J, there exists an index $\kappa \geq \iota, \iota'$ (take the union of ι and ι' in I) and transition homomorphisms

$$G/\mathrm{Ker}(\pi_\iota) \longrightarrow G/\mathrm{Ker}(\pi_\kappa) \quad \text{(canonical projection)}$$

since $\mathrm{Ker}(\pi_\iota) \supset \mathrm{Ker}(\pi_\kappa)$. Each $G/\mathrm{Ker}(\pi_\iota)$ can be identified to the compact Lie group $\pi_\iota(G)$ (closed subgroup of some $U_{d_\iota}(\mathbb{C})$, $d_\iota = \dim \pi_\iota$). By construction, the continuous homomorphism

$$G \longrightarrow \varprojlim_J G/\mathrm{Ker}(\pi_\iota) = \varprojlim_J \pi_\iota(G) \qquad \text{(canonical)}$$

is *one-to-one onto*. Since both groups are compact, it is also a topological homeomorphism. q.e.d.

Since Card(J) = Card(I) in the above construction, we see that if G has a countable basis for open sets, we can take I (and also J) to be countable so that G can be embedded in a countable product of Lie groups. This shows that G is metrizable (this application was already alluded to in sec.1, p.5). One should also point out that each $\pi_\iota(G) \cong G/\mathrm{Ker}(\pi_\iota)$ can very well be a *finite* group (finite groups are Lie groups of a very particular type...) in which case G is a *profinite group*. The reader will certainly have understood that we can obtain many variations on this theme...

Let us come back now to a more elementary proof of corollary (4.2) above. Start with a continuous function $f \in C(G)$ and a positive $\varepsilon > 0$. Choose an open symmetric neighbourhood U of the neutral element of G such that

$$y^{-1}x \in U \implies |f(y) - f(x)| < \varepsilon$$

(f is uniformly continuous on the compact group G). Take then a continuous function ψ on G such that

$$\psi \geqslant 0 \,, \quad \psi = \check{\psi} \,, \quad \mathrm{Supp}(\psi) \subset U \,, \quad \int_G \psi(y)\, dy = 1 \,.$$

Thus we can write

$$f(x) = f(x) \int_G \psi(y^{-1}x)\, dy$$

and coming back to the operator $K : L^2(G) \longrightarrow L^2(G)$ with kernel $k(x,y) = \psi(y^{-1}x)$, we see that

$$Kf(x) = \int_G f(y)\, \psi(y^{-1}x)\, dy \quad .$$

Thus

$$|Kf(x) - f(x)| = \int_G |f(y) - f(x)|\, \psi(y^{-1}x)\, dy <$$

$$< \int_U \varepsilon \cdot \psi(y^{-1}x)\, dy = \varepsilon \,.$$

On the other hand, the L^2-expansion

$$f = f_o + \sum f_i \qquad \text{(convergence in quadratic mean)}$$

with some continuous functions f_i $(i \neq 0)$ associated to the non-zero eigenvalues λ_i of K , leads to a uniformly convergent expansion

$$Kf = \sum \lambda_i f_i \qquad (f_o \in \mathrm{Ker}\, f \subset L^2(G))$$

(recall that K is even a compact operator from $L^2(G)$ into $C(G)$). Taking a sufficiently large finite partial sum, we shall have

$$\| Kf - \sum_{i \leqslant N} \lambda_i f_i \| < \varepsilon \qquad \text{(uniform norm).}$$

Summing up, we shall have $\| f - \sum_{i \leqslant N} \lambda_i f_i \| < 2\varepsilon$. It only remains to prove that the functions f_i are finite linear combinations of coefficients of (finite dimensional) irreducible representations. This follows from the next lemma.

(4.6) Lemma. Let $V \subset L^2(G)$ be a finite dimensional subspace which is invariant under the left regular representation of G. Then V consists of (classes of) continuous functions and each $f \in V$ can be written

$$f(x) = \mathrm{Tr}(A\,\overline{\pi}(x)) = \mathrm{Tr}(A\,\check{\overline{\pi}}(x)) \quad \text{for some} \quad A \in \mathrm{End}(V)$$

(here π denotes the restriction of the left regular representation to V).

Proof. Let us take an orthonormal basis (χ_i) of V and introduce the coefficients c^i_j of π. They are defined by

$$\pi(x)\,\chi_i = \sum_j c^j_i(x)\,\chi_j \qquad (x \in G).$$

If $f = \sum a^i \chi_i$, we have

$$\pi(x)f = \sum a^i\,\pi(x)\,\chi_i = \sum a^i c^j_i(x)\,\chi_j$$

hence

$$f(x) = (\ell(x^{-1})f)(e) = \sum c^j_i(x^{-1})\,a^i_j \qquad (\text{with } a^i_j = a^i\,\chi_j(e)).$$

Since $\pi(x^{-1}) = \pi(x)^*$, we have

$$c^j_i(x^{-1}) = \overline{c^i_j(x)} \quad \text{and} \quad f(x) = \mathrm{Tr}(A\,\overline{\pi}(x))$$

as claimed (cf. ex.2 of sec.2, p.20).

The reader will check that if π denotes now the restriction of the *right* regular representation on a finite dimensional invariant subspace $V \subset L^2(G)$, then any $f \in V$ can be written

$$f(x) = \mathrm{Tr}(A\,\pi(x)) \qquad (\text{for some } A \in \mathrm{End}(V)).$$

For this reason, we shall often prefer to work with the right regular representation.

EXERCISES

1. *Let G be a compact totally discontinuous group. Show that* A_G *is the algebra of all locally constant functions on G. (Observe that a locally constant function on G is uniformly locally constant, hence can be identified with a function on a quotient G/H where H is some open subgroup of G. Conversely, any finite dimensional representation of G must be trivial on an open subgroup H of G.)*

2. *Let G be any compact group. Show that* A_G *consists of the continuous functions f on G for which the left and right translates of f generate a finite dimensional subspace of C(G). In particular, if* G_1 *and* G_2 *are two compact groups, any continuous homomorphism* $h : G_1 \longrightarrow G_2$ *has a transpose* ${}^t h : A_2 \longrightarrow A_1$ $(A_i = A_{G_i})$ *defined by* ${}^t h(f) = f \circ h$ *(a priori this transpose is a linear mapping*

$${}^t h \ : \ C(G_2) \longrightarrow C(G_1) \quad).$$

3. *Let* $G = U_n(\mathbb{C})$ *with its canonical representation* π *in* $V = \mathbb{C}^n$. *Since* π *is unitary, we can identify* $\bar{\pi}$ *with the contragredient of* π*: it acts in the dual* V^* *of V.*

a) Let A^p_q *denote the space of linear combinations of coefficients of the representation*

$$\pi^p_q = \bar{\pi}^{\otimes p} \otimes \pi^{\otimes q} \quad \text{in} \quad (V^*)^{\otimes p} \otimes V^{\otimes q} = T^p_q(V) .$$

Prove that the sum of the subspaces A^p_q *of C(G) is an algebra A (show that* $A^p_q A^r_s \subset A^{pr}_{qs}$ *) stable under conjugation (show that* $\overline{A^p_q} = A^q_p$ *) which separates the points of G. Using the Stone-Weierstrass theorem, conclude that A is dense in C(G).*

b) Show that $A = A_G$ *(use part a to prove that any irreducible representation of G appears as subrepresentation of some* π^p_q *, or, in other words, can be realized on a space of mixed tensors).*

4. *Let G be a closed subgroup of* $U_n(\mathbb{C})$. *Using the fact that any finite dimensional representation of G appears in the restriction of some finite dimensional representation of* $U_n(\mathbb{C})$ *(this is a consequence*

of the theory of induced representations: cf. sec. 8), show that G is a real algebraic subvariety of $U_n(\mathbb{C})$ *. (The transpose of the embedding* $G \hookrightarrow U_n(\mathbb{C})$ *is the operation of restriction on polynomial functions and is surjective. Hence* A_G *is a quotient of the polynomial algebra A of* $U_n(\mathbb{C})$*. By the exercise 3, A is generated by the coordinate functions*

$$U_n(\mathbb{C}) \longrightarrow \mathbb{C}, \quad x = (x^i_j) \longmapsto x^i_j$$

and their conjugates.)

5 DECOMPOSITION OF THE REGULAR REPRESENTATION

In this section, we establish a connection between finite dimensional representations of G and subrepresentations of $L^2(G)$. Eventually, we shall be able to prove that any irreducible representation of a compact group is finite dimensional. The possibility of constructing a link between *any finite dimensional representation* and a subrepresentation of $L^2(G)$ is suggested by lemma (4.6). To avoid complex conjugations, we work with the *right regular representation.*

Let (π, V) be any finite dimensional representation of the compact group G. For any endomorphisms $A \in \mathrm{End}(V)$, we define the corresponding coefficient c_A of π by $c_A(x) = \mathrm{Tr}(A \cdot \pi(x))$. The right translates of these coefficients are easily identified

$$(\rho(s)\, c_A)(x) = c_A(xs) = \mathrm{Tr}(A\pi(x)\pi(s)) =$$

$$= \mathrm{Tr}(\pi(s) \cdot A\, \pi(x)) = \mathrm{Tr}(B\, \pi(x)) = c_B(x) \qquad (B = \pi(s) \cdot A).$$

Thus we consider the representation of G in End(V) defined by

$$\ell_\pi(s)(A) = \pi(s) \cdot A \qquad (s \in G\, , A \in \mathrm{End}(V)).$$

Then the above computation shows that $A \longmapsto c_A$ is a G-morphism

$$c : \mathrm{End}(V) \longrightarrow C(G) \subset L^2(G)$$

(intertwining ℓ_π and π).

Let us study this morphism more systematically. By lemma (4.6) we already know that its image contains all subrepresentations of $(L^2(G), \rho)$ which are equivalent to π. The situation is, of course, much simpler when π is irreducible, and we shall make this assumption from now on.

(5.1) Definition. Let π be a finite dimensional irreducible representation of a compact group G. We denote by $L^2(G, \pi)$ the sum of all subspaces of the right regular representation which are equivalent to π.

According to lemma (4.6), the subspace $L^2(G, \pi)$ is finite dimensional (and of dimension $\leqslant n^2$ if $n = \dim(V)$). A function $f \in L^2(G)$

belongs to this subspace $L^2(G,\pi)$ precisely when the right translates of f generate an invariant subspace (of the right regular representation) equivalent to a finite multiple of π (a finite sum of subrepresentations equivalent to π). We say that $L^2(G,\pi)$ is the *isotypical component* of π in $L^2(G)$. Obviously, this subspace $L^2(G,\pi)$ only depends on the equivalence class of π. We shall now prove that the dimension of an isotypical component $L^2(G,\pi)$ is exactly $(\dim\pi)^2$ (this result is quite similar to the finite groups case, but since $L^2(G)$ is infinite dimensional when G is infinite, we see that infinite compact groups *should have infinitely many classes of finite dimensional irreducible representations*... in order to exhaust $L^2(G)$). There are two methods for doing this: the first one is based on a purely algebraic lemma, while the second is analytical. As they both have important applications, we present them in succession.

(5.2) Burnside's theorem. Let G be any group and π any irreducible representation of G in a finite dimensional vector space V over an algebraically closed field k. Then

a) all operators $A \in \mathrm{End}(V)$ are (finite) linear combinations of the operators $\pi(s)$ $(s \in G)$,

b) for an operator $A \in \mathrm{End}(V)$, the relations $\mathrm{Tr}(A\,\pi(s)) = 0$ (for all $s \in G$) imply $A = 0$.

Proof. Let us denote by $\mathcal{U}$ the *enveloping algebra* of the representation π, namely, the space generated by all $\pi(s)$ $(s \in G)$ in End(V). Also define

$$\mathfrak{A} = \mathcal{U}^{\perp} = \{A \in \mathrm{End}(V) : \mathrm{Tr}(BA) = 0 \text{ for all } B \in \mathcal{U}\}.$$

Since $B \in \mathcal{U}$ implies $B' = B\cdot\pi(s) \in \mathcal{U}$ $(s \in G)$, we have

$$\mathrm{Tr}(B\,\pi(s)\,A) = \mathrm{Tr}(B'A) = 0 \quad \text{for } A \in \mathfrak{A}.$$

As above, let us consider the representation

$$\ell_\pi : G \to \mathrm{Gl}(\mathfrak{A}),\quad \ell_\pi(s)(A) = \pi(s)\cdot A .$$

Evaluations at elements $a \in V$

$$\alpha : \mathfrak{A} \to V,\quad A \mapsto Aa$$

are thus G-morphisms :

$$\alpha(\ell_\pi(s)(A)) = \alpha(\pi(s)\cdot A) = (\pi(s)\cdot A)(a) =$$

$$= \pi(s)(Aa) = \pi(s)(\alpha(A)) .$$

Let us show that $\mathfrak{A} \neq 0$ leads to a contradiction. In this case, we could take an invariant irreducible subspace $\mathfrak{A}_o \subset \mathfrak{A}$ (since we are working over a field of arbitrary characteristic, we cannot invoke complete reducibility, but we just take $\mathfrak{A}_o$ G-invariant, minimal among the non-zero subspaces of $\mathfrak{A}$). Take then $0 \neq A \in \mathfrak{A}_o$ and $a \in V$ with $Aa \neq 0$. The corresponding evaluation operator α is not zero and is a G-morphism from $(\mathfrak{A}_o, \ell_\pi)$ to (V, π). This implies that α is an isomorphism ($\mathrm{Ker}(\alpha)$ and $\mathrm{Im}(\alpha)$ must be invariant subspaces, and $\mathrm{Ker}(\alpha) \neq V$, $\mathrm{Im}(\alpha) \neq 0$). Let us take a k-basis (e_i) of V with $e_1 = a$ and denote by α_i the corresponding evaluation operators. Then the operators $\alpha_1^{-1}\alpha_i$ must commute to the irreducible G-action in $\mathfrak{A}_o$. By Schur's lemma, they must be scalar operators

$$\alpha_i = \lambda_i \alpha_1 \qquad (\lambda_i \in k,\ \lambda_1 = 1).$$

To be able to compute traces, we introduce the dual basis (ε_i) of (e_i) (it is defined by $\varepsilon_i(e_j) = \delta_{ij} = 0$ if $i \neq j$ and 1 if $i = j$).
With our choices $(e_1 = a,\ Aa \neq 0)$, we have

$$0 = \mathrm{Tr}(\pi(s)A) = \sum\langle \varepsilon_i, \pi(s)Ae_i\rangle = \sum\langle\varepsilon_i, \pi(s)\alpha_i(A)\rangle =$$
$$= \sum\langle\varepsilon_i, \lambda_i\pi(s)\alpha_1(A)\rangle = \langle\sum\lambda_i\varepsilon_i, \pi(s)Aa\rangle \qquad (s \in G).$$

But $Aa \neq 0$ implies that the vectors $\pi(s)Aa$ generate V (this space is irreducible under G!) and we infer

$$\sum\lambda_i\varepsilon_i = 0, \quad \lambda_i = 0$$

contradicting $\lambda_1 = 1$! q.e.d.

(5.3) <u>Corollary</u>. <u>Let π be a finite dimensional irreducible representation of G (in V). Then, the representation $\sigma \times \tau$ of $G \times G$ in End(V) defined by</u>

$$\sigma\times\tau\,(s,t)\,(A) = \pi(s)\cdot A\cdot\pi(t)^{-1}$$

<u>is irreducible</u>.

<u>Proof</u>. Let W denote the $G \times G$-module End(V) and take any $G \times G$-morphism $\Phi : W \longrightarrow W$. We have

$$\Phi(\pi(g)A) = \pi(g)\,\Phi(A) \quad \text{hence} \quad \Phi(\pi(g)) = \pi(g)\,\Phi(I)$$

whence

$$\Phi(A) = A\cdot\Phi(I) \quad \text{for } all\ A \in W = \mathrm{End}(V)$$

(by the theorem). But we must also have $\Phi(\pi(g)) = \Phi(I)\,\pi(g)$ and thus the operator $\Phi(I)$ commutes to all operators $\pi(g)$ $(g \in G)$. Since π is irreducible, Schur's lemma again implies that $\Phi(I) = \lambda I$ is a scalar operator. Consequently, as we have seen, $\Phi(A) = \lambda A$. In particular, there is no projector in W other than 0 and 1 which is a $G \times G$-morphism : there are no invariant subspaces in W other than 0 and W ($\neq 0$) itself. The corollary is thus completely proved.

The fact that the G-morphism

$$c \ : \ \mathrm{End}(V\ ) \longrightarrow L^2(G, \pi) \quad (\subset C(G) \subset L^2(G)\ )$$
$$A \longmapsto c_A$$

is injective (hence an isomorphism on the isotypical component of π) follows from Burnside's theorem (5.2.b). Alternatively, one can check that c is a $G \times G$-morphism : using the notations of the corollary ($\ell_\pi = \sigma$, $\rho_\pi = \tau$) we indeed have

$$c_A(s^{-1}xt) = \mathrm{Tr}(A\,\pi(s^{-1})\,\pi(x)\,\pi(t)) =$$
$$= \mathrm{Tr}(\,\pi(t)\,A\,\pi(s)^{-1}\pi(x)) = c_{\sigma \times \tau(t,s)}(x) \quad .$$

Since $c \neq 0$ and End(V) is irreducible under $\sigma \times \tau$, c must be an isomorphism into $L^2(G)$.

It is interesting to observe that Burnside's theorem admits a natural generalization for $k = \mathbb{C}$ and π *unitary* (in any Hilbert space). The density theorem of von Neumann asserts that if π is any unitary (topologically) irreducible representation of *any group* G, then the set of operators $\pi(g)$ $(g \in G)$ is dense in $\mathcal{L}(H) = \mathrm{End}_{\mathrm{cont.}}(H)$ for the strong, or the weak topology on this set of operators.

Let us turn now to the analytic study of the G-morphism $c : \mathrm{End}(V) \longrightarrow L^2(G)$, $A \longmapsto c_A = \mathrm{Tr}(A\pi)$. The fact that $c_A \neq 0$ for $A \neq 0$ can be deduced from a computation of the quadratic norm of these coefficient functions. It is easier to start with the case of rank $\leqslant 1$ linear mappings. To bring these special maps into the limelight, we use the isomorphism

$$V^\vee \otimes V \longrightarrow \mathrm{End}(V) \qquad (V^\vee = \text{dual of } V)$$

defined as follows. If $u \in V^\vee$ and $v \in V$, the operator (corresponding to) $u \otimes v$ is

$$u \otimes v \ : \ x \longmapsto u(x)\,v = \langle u,x\rangle\, v \quad .$$

The image of $u \otimes v$ consists of multiples of v and $u \otimes v$ has rank 1 when u and v are non-zero (quite generally, decomposable tensors correspond to operators of rank $\leqslant 1$). The coefficient c_A with respect to the operator $A = u \otimes v$ coïncides with the previously defined coefficient

$$c_v^u = \langle u, \pi(x)\, v\rangle = c_{u \otimes v}(x)$$

(cf. first exercise at the end of this section).

(5.4) *Fundamental lemma*. *Let* π *and* σ *be two representations of a compact group* G *and* $A : V_\pi \longrightarrow V_\sigma$ *be a linear mapping*. *Then*

$$A^\natural = \int_G \sigma(g)\, A\, \pi(g)^{-1}\, dg$$

is a G-*morphism from* V_π *to* V_σ : $A^\natural \in \mathrm{Hom}_G(V_\pi, V_\sigma)$.

The *proof* of this lemma is solely based on the invariance of the Haar measure :

$$A^\natural \pi(s) = \int \sigma(g)\, A\, \pi(g)^{-1} \pi(s)\, dg = \int \sigma(g)\, A\, \pi(s^{-1} g)^{-1}\, dg$$

and replacing g by sg (i.e. $s^{-1}g$ by g)

$$A^\natural \pi(s) = \int \sigma(sg)\, A\, \pi(g)^{-1}\, dg = \sigma(s)\, A^\natural \quad . \qquad \text{q.e.d.}$$

Thus the averaging operation (given by the Haar integral) of lemma (5.4) leads to a *projector*

$$\natural : \mathrm{Hom}(V_\pi, V_\sigma) \longrightarrow \mathrm{Hom}_G(V_\pi, V_\sigma) \quad , \quad A \longmapsto A^\natural$$

(this point of view is expanded in the second exercise at the end of this section, where an alternative proof of (5.4) can be found). In particular, when π and σ are *disjoint*, i.e. $\mathrm{Hom}_G(V_\pi, V_\sigma) = 0$, we must have $A^\natural = 0$! This is certainly the case if π and σ are non-equivalent irreducible representations (Schur's lemma). Another case of special interest is $\pi = \sigma$ finite dimensional and irreducible. Schur's lemma gives $\mathrm{Hom}_G(V_\pi, V_\sigma) = \mathbb{C}$ and thus $A = \lambda_A\, \mathrm{id.}$ is a scalar operator. It is not difficult to give the value of λ_A as a function of A in this case.

(5.5) *Proposition*. *If* π *is a finite dimensional irreducible representation of the compact group* G *in* V, *the projector*

$$\mathrm{End}(V) \longrightarrow \mathrm{End}_G(V) = \mathbb{C}\ \mathrm{id.} \quad , \quad A \longmapsto A^\natural = \lambda_A \mathrm{id.}$$

is given explicitly by the following formula :

$$A^{\natural} = \int_G \pi(g)\, A\, \pi(g)^{-1}\, dg = \frac{\mathrm{Tr}(A)}{\dim V}\, \mathrm{id}_V \quad .$$

Proof. Since we know a priori that the operator $A^{\natural}$ is a scalar operator $\lambda_A \mathrm{id}.$, we can determine the value of the scalar λ_A simply by taking traces in the defining equalities

$$\lambda_A \mathrm{Tr}(\mathrm{id}_V) = \mathrm{Tr} \int_G \cdots = \int_G \mathrm{Tr}(\pi(g)\, A\, \pi(g)^{-1})\, dg =$$

$$= \int_G \mathrm{Tr}(A)\, dg = \mathrm{Tr}(A) \quad . \qquad \text{q.e.d.}$$

(5.6) Theorem (Schur's orthogonality relations). Let G be a compact group and π, σ two finite dimensional irreducible representations of G. Assume π and σ unitary. Then

a) if π and σ are non-equivalent, $L^2(G, \pi)$ and $L^2(G, \sigma)$ are orthogonal in $L^2(G)$,

b) if π and σ are equivalent, $L^2(G, \pi) = L^2(G, \sigma)$ and the scalar product of two coefficients of this space is given by

$$(c^u_v \mid c^x_y) = \int_G \overline{(u \mid \pi(g)v)}\, (x \mid \pi(g)y)\, dg = \overline{(u \mid x)}\, (v \mid y)/\dim V ,$$

c) more generally in the case $\pi = \sigma$, the scalar product of general coefficients is given by

$$(c_A \mid c_B) = \int_G \overline{\mathrm{Tr}(A\pi(g))}\, \mathrm{Tr}(B\pi(g))\, dg = \mathrm{Tr}(A^* B)/\dim V \quad .$$

Proof. a) follows from the fundamental lemma (5.4) and b) follows similarly from the proposition (5.5). It will be enough to show how b) is derived. For this purpose, we consider the particular operators $\bar{v} \otimes y$ $(\bar{v} \in \bar{V}_\pi = \bar{V}$, $y \in V)$ and apply the result of the proposition

$$\int_G \pi(g)\, \bar{v} \otimes y\, \pi(g)^{-1}\, dg = \frac{\mathrm{Tr}(\bar{v} \otimes y)}{\dim V}\, \mathrm{id}_V = \frac{(v \mid y)}{\dim V}\, \mathrm{id}_V$$

(cf. ex.1 at the end of the section for the computation of the trace of the operator $\bar{v} \otimes y$) . Let us apply this operator to the vector u, and take the scalar product with the vector x

$$\left(x \,\Big|\, \int \pi(g)\, \bar{v} \otimes y\, \pi(g)^{-1}\, u\, dg\right) = \frac{(v \mid y)}{\dim V}\, (x \mid u) = \frac{\overline{(u \mid x)}\, (v \mid y)}{\dim V} \quad .$$

But we have

$$\pi(g)\ \bar{v} \otimes y\ \pi(g)^{-1} u = \pi(g)\ (v \mid \pi(g^{-1})\ u)\ y =$$

$$= (\pi(g)v \mid u)\ \pi(g)\ y = \overline{(u \mid \pi(g)\ v)}\ \pi(g)\ y$$

hence

$$(x \mid \int_G \ldots) = \int_G \overline{(u \mid \pi(g)\ v)}\ (x \mid \pi(g)\ y)\ dg = (c_v^u \mid c_y^x)$$

as expected. Finally, c) follows from b) by linearity since the operators $\bar{v} \otimes y$ (of rank $\leqslant 1$) generate End(V). q.e.d.

In particular we see that if $0 \neq A \in \mathrm{End}(V)$,

$$\|c_A\|^2 = \mathrm{Tr}(A^*A)/\dim V \neq 0$$

and the mapping $c : \mathrm{End}(V) \longrightarrow L^2(G,\pi)$ is one-to-one (onto). The dimension of this isotypical component is thus $(\dim V)^2$.

(5.7) Corollary. The Hilbert space $L^2(G)$ is the Hilbert sum of all isotypical components

$$L^2(G) = \widehat{\bigoplus}\ L^2(G,\pi)$$

(the summation index π runs over *equivalence classes* of finite dimensional irreducible representations of the compact group G).

Proof. We have already seen (in the theorem) that the isotypical subspaces $L^2(G,\pi)$ are mutually orthogonal to each other. Thus, our corollary will be proved if we show that the algebraic sum

$$A_G = \bigoplus L^2(G,\pi) \subset C(G)$$

is *dense* in the Hilbert space $L^2(G)$. But A_G consists of coefficients of finite dimensional representations of G (we have indeed proved that all finite dimensional representations are completely reducible), and the Peter-Weyl theorem (4.2) has shown that A_G is dense in C(G) for the uniform norm. A fortiori A_G will be dense in $L^2(G)$ for the quadratic norm. q.e.d.

(5.8) Corollary. Any (continuous, topologically) irreducible representation of a compact group G in a Banach (or barrelled) space is finite dimensional.

Proof. Let $\sigma : G \rightarrow Gl(E)$ be such a representation, and let E' denote the (topological) dual of E : E' is the Banach space (or locally convex space) of continuous linear forms on E (by the Hahn-Banach theorem, for each $0 \neq x \in E$, there is a continuous linear form $x' \in E'$

with $\langle x',x\rangle = x'(x) \neq 0$). For $u \in E'$ and $v \in E$, we can consider the corresponding coefficient of σ

$$c_v^u \in C(G) \quad : \quad g \mapsto c_v^u(g) = \langle u, \sigma(g)\, v\rangle .$$

Letting v vary in E, we get a linear mapping

$$Q \ : \ E \to C(G) \subset L^2(G) \ , \quad v \mapsto c_v^u \quad .$$

Since G is compact and the mappings $g \mapsto \sigma(g)\, v$ are continuous (by the definition of *continuous* representations), the sets $\sigma(G)v$ are compact hence bounded in E ($v \in E$). By the uniform boundedness principle (Banach-Steinhaus theorem), the set $\sigma(G)$ of operators in E is *equicontinuous* and bounded

$$\operatorname*{Sup}_{g\in G} \|\sigma(g)\| = M < \infty \quad .$$

Hence

$$\|Q v\| = \operatorname*{Sup}_{g\in G} |c_v^u(g)| = \operatorname*{Sup}_{G} |\langle u , \sigma(g)\, v\rangle|$$

$$\leqslant \|u\|_{E'} \operatorname*{Sup}_{G} \|\sigma(g)\, v\|_E \leqslant M \|u\|_{E'} \|v\|_E \quad .$$

This proves that Q is continuous from E into C(G) (equipped with the uniform norm): its kernel is a proper and closed subspace $F \neq E$ if $u \neq 0$ (in this case, $u(v) \neq 0$ for some $v \in E$ and thus $c_v^u(e) = \langle u,v\rangle = u(v) \neq 0$). Take $v \in E$ with $Q(v) \neq 0$, and apply the orthogonal Hilbert sum decomposition of the preceding corollary to Q(v)

$$\sum P_\pi(Q\, v) = Q v \neq 0$$

with

$$P_\pi = \text{orthogonal projector from } L^2(G) \text{ onto } L^2(G,\pi) \ .$$

This implies that there is a π (finite dimensional irreducible representation of G) with $P_\pi Q\, v \neq 0$. For this π, we consider the composite

$$E \xrightarrow{Q} L^2(G) \xrightarrow{P_\pi} L^2(G,\pi)$$

and its kernel which is a proper closed subspace $F_\pi \neq E$. But Q is a G-morphism (intertwinning σ and the right regular representation)

$$c^u_{\sigma(x)v}(g) = \langle u, \sigma(g)\sigma(x)v\rangle = c_v^u(gx)$$

$$\text{implies} \quad Q(\sigma(x)\, v) = \rho(x)\, Q(v) \quad .$$

Since P_π is also a G-morphism, the kernel F_π of the composite $P_\pi Q$ must be an invariant subspace of E . But E is irreducible by assumption, so that F_π must be $\{0\}$, and the composite $P_\pi Q$ is one-to-one (into):

$$\dim E \leqslant \dim L^2(G,\pi) = (\dim V)^2 \quad . \qquad \text{q.e.d.}$$

The preceding *finiteness* result can be proved independently from the theorem (not using its corollary (5.7)) (cf. sec.7 below for a proof in the category of *quasi-complete* vector spaces, and sec.8 for a direct proof in Hilbert spaces). In this second proof, we still use the intertwining operator

$$Q : E \longrightarrow L^2(G) , \quad v \longmapsto c^u_v$$

which must be one-to-one (into) when non-zero ($u \neq 0$) by irreducibility assumption on E. Let us show that E is a *strict morphism* (the inverse of E, defined on the image of E is also continuous) : with the above definition of M,

$$\|v\| = \|\sigma(g^{-1})\sigma(g)\, v\| \leqslant \|\sigma(g^{-1})\|\, \|\sigma(g)\, v\| \leqslant M\, \|\sigma(g)\, v\|$$

implies

$$\|\sigma(g)\, v\| \geqslant M^{-1}\, \|v\| \quad .$$

Since E is metrizable and complete (this proof works for *Frechet spaces*), Im(Q) = Q(E) will also be complete, hence *closed* in $L^2(G)$. Since the image of Q consists of continuous functions (and the Haar measure of a compact group is bounded), the conclusion follows from the next lemma.

(5.9) Lemma (Godement). Let X be a locally compact space, m a positive bounded measure on X, V a closed subspace of $L^2(X,m)$. If $V \subset L^\infty(X,m)$, then V is finite dimensional.

Godement's proof. Taking the *spectrum of the commutative von Neumann algebra* $L^\infty(X,m)$ (this is a compact *Stonian* space...), we are reduced to the case where V consists of continuous functions (anyway, this is the case in our application !). Replacing X by the support of m, we can assume moreover that

$$V \subset C_{\text{bounded}}(X) \hookrightarrow L^2(X,m) \quad \text{is one-to-one (into)} .$$

Since m is bounded, this embedding is continuous and V is *a fortiori* closed in $C_{\text{bounded}}(X)$ (equipped with the uniform norm). The Banach homomorphism theorem (Rudin 1973, p.47) asserts then that the uniform norm and quadratic norm are *equivalent on* V :

$$(*) \qquad \|f\| \leqslant c\|f\|_2 \qquad (f \in V) .$$

Hence evaluation forms (at points $x \in X$) are continuous on V equipped with the quadratic norm, and are given by scalar product (Riesz' theorem)

$$f(x) = \varepsilon_x(f) = (k_x \mid f) \qquad (f \in V) .$$

Taking an orthonormal basis (e_i) of V, we shall have

$$k_x = \sum (e_i \mid k_x)\, e_i = \sum \overline{e_i(x)}\, e_i$$

(with a convergence in $L^2(X,m)$). Parseval's identity gives

$$\|k_x\|^2 = \sum |e_i(x)|^2 \quad .$$

But (*) shows that it was possible to choose the vectors k_x with $\|k_x\| \leqslant c$ (anyway, the k_x make up a simply bounded set : they must be uniformly bounded...) and thus

$$\sum |e_i(x)|^2 \leqslant c^2 \quad .$$

The measurable function $x \mapsto \sum |e_i(x)|^2$ (Egorov's theorem is applicable) is bounded, hence integrable and we can write

$$\dim(V) = \sum \|e_i\|^2 = \sum \int |e_i(x)|^2 \, dm(x) =$$

$$= \int \sum |e_i(x)|^2 \, dm(x) \leqslant c^2 \, m(X) < \infty \quad .$$

Hörmander's proof. It still uses the inequality (*) written in the form

$$\text{Sup ess } \left|\sum c_i e_i(x)\right|^2 \leqslant c^2 \sum |c_i|^2$$

for any *finite* orthonormal system in V, hence

$$(**) \qquad \left|\sum c_i e_i(x)\right|^2 \leqslant c^2 \sum |c_i|^2$$

outside a negligible set $N(c_1,\ldots,c_n)$. The countable union N of the sets $N(c_1,\ldots,c_n)$ where the c_i (finite in number) have rational real and imaginary parts is still negligible and the above inequality is true outside N for all those finite families. As both sides of (**) are obviously continuous in the $c_i \in \mathbb{C}$, it will still hold for all c_i (but $x \notin N$). Take $c_i = \overline{e_i(x)}$! We obtain

$$\left(\sum |e_i(x)|^2\right)^2 \leqslant c^2 \sum |e_i(x)|^2$$

and

$$\sum |e_i(x)|^2 \leqslant c^2 \qquad (x \notin N).$$

Integrating, we find

$$n = \int \sum_{1\leqslant i\leqslant n} |e_i(x)|^2 \, dm(x) \leqslant c^2 \, m(X)$$

whence finally

$$\dim V = \operatorname{Sup} n \leqslant c^2 \, m(X) \quad . \qquad \text{q.e.d.}$$

Coming back to representation theory, we give a definition.

(5.10) <u>Definition</u>. <u>The dual</u> $\hat{G}$ <u>of a compact group</u> G <u>is the</u> *set* <u>of equivalence classes of irreducible representations of</u> G.

Let π be an irreducible representation of the compact group G, $\varpi = [\pi]$ its equivalence class ($\varpi \in \hat{G}$). We say that π is a *model* of ϖ in this case, i.e. when $\pi \in \varpi$. The dimension of ϖ is defined as $\dim(\pi) = \dim(V_\pi)$ independently from the model π chosen. Similarly, the isotypical component of ϖ (in the right regular representation) is defined by $L^2(G,\varpi) = L^2(G,\pi)$ independently from the model π chosen for ϖ. By the finiteness of the dimension of the irreducible representations of a compact group (5.8) and (5.7), we can write

$$L^2(G) = \widehat{\bigoplus_{\varpi\in\hat{G}}} L^2(G,\varpi) \quad .$$

In this way, the dual $\hat{G}$ of G is in one-to-one correspondence with a subset of closed subspaces (even finite dimensional subspaces) of $L^2(G)$. In particular, this dual $\hat{G}$ is a *set*... (mathematicians only rarely consider such trivialities !). Instead of $\pi \in \varpi \in \hat{G}$ we shall usually write more simply $\pi \in \hat{G}$ (hoping that this notation will not lead to any confusion).

(5.11) <u>Proposition</u>. <u>Let</u> G <u>be a compact group</u>. <u>Then the following properties are equivalent</u>.

i) <u>The dual</u> $\hat{G}$ <u>is countable</u>.

ii) $L^2(G)$ <u>is separable</u> (<u>countable Hilbert dimension</u>).

iii) G <u>is metrizable</u>.

<u>Proof</u>. The equivalence between i) and ii) is obvious since $L^2(G)$ is the Hilbert sum of the finite dimensional isotypical components $L^2(G,\varpi)$ over the index set $\hat{G}$. Moreover, G can always be embedded in a product

$$\prod_{\hat{G}} U(\pi) \quad \text{with } U(\pi) \cong U_{\dim(\pi)}(\mathbb{C}) \text{ (metrizable group).}$$

Since any countable product of metrizable topological spaces is metrizable, we see that i) $\Longrightarrow$ iii). Finally, the implication iii) $\Longrightarrow$ ii) is a classical application of the Stone-Weierstrass theorem (group theory is not involved in this point, cf. Dieudonné 1960, (7.3.1) p.131 and (7.4.4) p.134).

EXERCISES

1. Let V be a finite dimensional vector space, $V^{\vee}$ its dual and for $u \in V^{\vee}$, $v \in V$ denote by $u \otimes v$ the operator $x \longmapsto u(x)\,v$ as defined in this section. Show

a) $Tr(u \otimes v) = \langle u,v \rangle = u(v)$ *(intrinsic definition of Tr),*

b) $(u \otimes v)\circ(x \otimes y) = \langle u,y \rangle\, x \otimes v$,

c) ${}^{t}A\,u \otimes B\,v = B\circ(u \otimes v)\circ A$ *for* $A,\ B \in End(V)$.

Moreover, identifying the dual of $V^{\vee} \otimes V$ to $V \otimes V^{\vee}$ (in the obvious canonical way !), show

d) ${}^{t}(u \otimes v) = v \otimes u$.

Assume now that V is a representation space for a group G . Using a) and c) prove

e) $c_{u \otimes v} = c_{v}^{u}$ *(i.e.* $Tr(\pi(x)\circ u \otimes v)) = \langle u, \pi(x)\,v \rangle$ *).*

2. Let G be a compact group and (π, V) be a finite dimensional representation of G. We denote by V^{G} the subspace of invariants of V :

$$V^{G} = \{ v \in V : \pi(g)\,v = v \text{ for all } g \in G \} .$$

a) Check that the operator $P = \int_{G} \pi(g)\,dg$ is a projector from V onto V^{G}. If π is unitary, P is the orthogonal projector on this subspace.

b) For two finite dimensional representations π and σ of G, consider $Hom(V_{\pi}, V_{\sigma})$ as representation space of G via the action $g \cdot A = \pi(g)\circ A \circ \sigma(g)^{-1}$. Observe that

$$Hom(V_{\pi}, V_{\sigma})^{G} = Hom_{G}(V_{\pi}, V_{\sigma})$$

and deduce a proof of the fundamental lemma (5.4) from this observation.

c) Using the G-isomorphism

$$V_\pi^\vee \otimes V_\sigma \longrightarrow \mathrm{Hom}(V_\pi, V_\sigma)$$

($V_\pi^\vee \otimes V_\sigma$ being equipped with the representation $\pi^\vee \otimes \sigma$), conclude that the projector $\natural : V_\pi^\vee \otimes V_\sigma \longrightarrow (V_\pi^\vee \otimes V_\sigma)^G$ *is given by* $\int_G (\pi^\vee \otimes \sigma)(g)\, dg$.

3. Rewrite Schur's orthogonality relations (5.6.b) without assuming that π *is unitary.* *Hint: Compute the integral*

$$\int_G c_v^u(g^{-1})\, c_y^x(g)\, dg$$

(observe that

$$c_v^u(g^{-1}) = \check{c}_v^u(g) \text{ , } \check{c}_v^u \text{ is the } c_u^v\text{-coefficient of } \check{\pi} \text{) .}$$

6 CONVOLUTION, PLANCHEREL FORMULA & FOURIER INVERSION

In this section, we gather up odds and ends.

CONVOLUTION

On a compact group G, the *convolution* of two continuous functions f and g is defined by

$$f * g\,(x) = \int_G f(y)\, g(y^{-1}x)\, dy \quad .$$

Defining $f^*(x) = \overline{f(x^{-1})}$, we can also write

$$f * g\,(x) = \int f(xy)\, g(y^{-1})\, dy = \int f(xy^{-1})\, g(y)\, dy =$$

$$= \int \overline{f^*(yx^{-1})}\, g(y)\, dy = \int (\rho(x^{-1}) f^* \mid g) \quad .$$

The Cauchy-Schwarz inequality gives thus

$$|f * g\,(x)| \leqslant \|f^*\|_2 \, \|g\|_2 = \|f\|_2 \|g\|_2$$

whence

$$\|f * g\|_\infty = \operatorname*{Sup}_{x \in G} |f * g\,(x)| \leqslant \|f\|_2 \|g\|_2 \quad .$$

Consequently, the convolution product can be extended by continuity from C(G) to $L^2(G)$ and by definition,

$$* \; : \; L^2(G) \times L^2(G) \longrightarrow C(G) \; , \quad (f,g) \longmapsto f * g$$

is a continuous bilinear mapping (still satisfying the above inequality). On the other hand, the preceding formulas show

$$(f \mid g) = f^* * g\,(e) \; , \quad \|f\|_2^2 = f^* * f\,(e) \quad .$$

Since the group G is compact, the constant 1 is in $L^2(G)$ and for any $f \in L^2(G)$, the product $f = f \cdot 1$ must be in $L^1(G)$: $L^2(G) \subset L^1(G)$. The convolution product can even be defined in $L^1(G)$ by the same integral formula, but this integral will not converge for every $x \in G$ and the result will no more be continuous. To see what happens, let us take f and g in $L^1(G)$. I claim that $x \longmapsto f * g\,(x) = \int f(y)\, g(y^{-1}x)\, dy$ is also an integrable function. We apply Fubini's theorem as follows

$$|f * g(x)|\,dx \leqslant \int dx \int dy\ |f(y)\, g(y^{-1}x)| =$$
$$= \int dy\,|f(y)| \int dx\,|g(y^{-1}x)| = \|g\|_1 \int |f(y)|\,dy =$$
$$= \|f\|_1\, \|g\|_1 < \infty \quad .$$

These inequalities prove that $\int f(y)\, g(y^{-1}x)\, dy$ converges absolutely for nearly all x in G (all functions considered are *measurable*...) and the result $f * g \in L^1(G)$ satisfies

$$\|f * g\|_1 \leqslant \|f\|_1 \|g\|_1 \quad .$$

This convolution product is associative (cf. infra.) and $L^1(G)$ is an *algebra for convolution*. This algebra has no unit element in general (more precisely, it has no unit when G is not discrete, i.e. G not finite). We shall also see that this algebra $L^1(G)$ is commutative exactly when G is commutative.

INTEGRATION OF REPRESENTATIONS

Let us assume first that π is a *unitary* representation of the compact group G in a Hilbert space H. We can "extend" π to a representation of $L^1(G)$ by the formula

$$\pi^1(f) = \int_G f(x)\pi(x)\,dx \qquad (f \in L^1(G)).$$

These integrals converge absolutely in norm : $\|\pi(x)\| = 1$ implies

$$\int \|f(x)\,\pi(x)\|\,dx = \int |f(x)|\,dx = \|f\|_1 \quad .$$

Thus we even have

$$\|\pi^1(f)\| \leqslant \|f\|_1 \qquad (f \in L^1(G)).$$

Although G is not really embedded in $L^1(G)$ (when G is infinite), we consider π^1 as an extension of π, as in the finite group case. Later on we shall even drop the index 1, writing π instead of π^1. The association

$$\pi : G \longrightarrow Gl(H) \quad \rightsquigarrow \quad \pi^1 : L^1(G) \longrightarrow End(H)$$

can even be made when π is a representation in a Banach space since $\pi(G)$ is always a bounded set of End(H) : being weakly compact, it is uniformly bounded.

Let me introduce here some <u>comments on vector integration</u>. The definition of $\dot{\pi}^1$ is obtained by integration of the vector functions

$$G \longrightarrow \mathrm{End}(H), \quad g \longmapsto f(g)\pi(g) \qquad (f \in L^1(G)).$$

Thus our definition was based on a *strong* integral (converging in norm) defined for functions with values in a Banach space. This theory is expounded for example in Bourbaki (Integration, 1965, Chap.II) and uses for instance the following approximation lemma

> if f is continuous with compact support and takes its values in a Banach space, for every $\varepsilon > 0$, there exists a partition of unity (φ_i) with continuous functions φ_i and suitable points $x_i \in \mathrm{Supp}(\varphi_i)$ with
>
> $$\| f(x) - \sum_{\text{finite}} \varphi_i(x) f(x_i) \| \leqslant \varepsilon \qquad \text{(loc. cit. p.44)} .$$

There is another notion of *weak* integral, extending the preceding one (loc. cit. Chap.III). If $f : G \longrightarrow E$ (locally convex space) is weakly continuous with compact support, the integrals

$$\int \langle x', f(x) \rangle \, dx$$

are well defined for each linear form $x' \in E'$. Obviously

$$x' \longmapsto \int \langle x', f(x) \rangle \, dx$$

is a linear form on E' (this linear form is not necessarily continuous on E') hence an element of the algebraic dual $(E')^*$ of E'. This element of $(E')^*$ is simply denoted $\int f(x)\,dx$. From its definition as linear form, we have

$$\langle x', \int f(x)\,dx \rangle = \int \langle x', f(x) \rangle \, dx \quad .$$

A general theorem about (locally convex topological vector spaces) asserts that $(E')^*$ can be canonically identified with the $\sigma(E,E')$-weak completion of E. But when E is *quasi-complete* (i.e. each closed bounded set in E is complete), one proves easily that

$$\int f(x)\,dx \in E \subset (E')^* \qquad (f \in C(G))$$

(loc. cit. Prop.7,p.79).

For this reason, all results obtained by integration of representations are valid for representations in *quasi-complete* topological vector spaces ...

COMPARISON OF SEVERAL NORMS

Let us start by recalling the definition of the Hilbert-Schmidt norm of operators in finite dimensional Hilbert spaces V (for general Hilbert spaces, cf. sec.8 below). Let $A \in \mathrm{End}(V)$ be any endomorphism in V. Take any orthonormal basis (e_i) of V and assume that A is represented by the matrix (a^i_j) in the basis (e_i). Obviously

$$\|A\|_2^2 = \sum_{i,j} |a^i_j|^2$$

defines a norm on End(V) (a priori, this norm depends on the choice of orthonormal basis (e_i)). If B is another endomorphism, represented by the matrix (b^i_j) (in the same basis), a small computation shows that (matrices are to be multiplied line by column)

$$\mathrm{Tr}(A^*B) = \sum_{i,j} \bar{a}^i_j \, b^i_j \quad .$$

This shows that the Hilbert-Schmidt norm is derived from the Hilbert-Schmidt scalar product

$$(A \mid B)_2 = \sum_{i,j} \bar{a}^i_j \, b^i_j = \mathrm{Tr}(A^*B)$$

on End(V), and is in particular independent from the choice of orthonormal basis (e_i) of V.

Now, we come back to a compact group G and a unitary irreducible representation $\pi \in \hat{G}$ in some finite dimensional Hilbert space $V = V_\pi$. The spaces

$$V^\vee \otimes V \ , \quad \mathrm{End}(V) \ , \quad L^2(G,\pi)$$

are G-isomorphic. We shall give explicit isomorphisms between them, keeping track of the various norms involved. We have introduced the coefficients of

$$c^u_v(x) = \langle u, \pi(x)v \rangle \qquad (u \in V^\vee,\ v \in V)\ ,$$

and the more general coefficients (linear combinations of the preceding ones)

$$c_A(x) = \mathrm{Tr}(A\pi(x)) \qquad (A \in \mathrm{End}(V)\ )$$

with the relation

$$c_A = c^u_v \quad \text{for} \quad A = u \otimes v \quad .$$

For fixed $u \in V$, the linear mapping

$$c^u : V \longrightarrow L^2(G,\pi) , \quad v \longmapsto c_v^u$$

is a G-morphism from π to ρ (right regular representation). Similarly, if $v \in V$ is fixed,

$$(\ell(s)c_v^u)(x) = c_v^u(s^{-1}x) = \langle u, \pi(s^{-1})\pi(x)v \rangle =$$

$$= \langle {}^t\pi(s^{-1})u, \pi(x)v \rangle = c_v^{\check{\pi}(s)u}(x)$$

shows that the linear mapping

$$c_v : V^{\vee} \longrightarrow L^2(G,\pi) , \quad u \longmapsto c_v^u$$

is a G-morphism from $\check{\pi}$ to ℓ (left regular representation). Summing up,

$$c : V^{\vee} \otimes V \longrightarrow L^2(G,\pi) , \quad u \otimes v \longmapsto c_v^u$$

is a $G \times G$-morphism from $\check{\pi} \otimes \pi$ to $\ell \times \rho$ (biregular regular representation). The action of $G \times G$ on End(V) (cf. sec.5) is defined by

$$(\ell \times \rho(s,t))c_A(x) = c_A(s^{-1}xt) = \mathrm{Tr}(A\pi(s^{-1})\pi(x)\pi(t)) =$$

$$= \mathrm{Tr}(\pi(t)\, A\, \pi(s)^{-1}\pi(x)) = c_{\pi(t)\, A\, \pi(s)^{-1}}(x) ,$$

hence is given by

$$(s,t) \cdot A = \pi(t)\, A\, \pi(s)^{-1} .$$

In the following diagram of G-morphisms, $V^{\vee} \otimes V$ is equipped with the scalar product defined by

$$(u \otimes v \mid x \otimes y) = (u \mid x)^{\vee} (v \mid y) = \overline{(u \mid x)}(v \mid y)$$

(we still use the Riesz' isomorphism between $V^{\vee}$ and $\bar{V}$). This scalar product corresponds to the Hilbert-Schmidt norm

$$(u \otimes v \mid x \otimes y)_2 = \mathrm{Tr}((u \otimes v)^* x \otimes y) .$$

Here is a <u>diagram</u> summing up the whole situation.

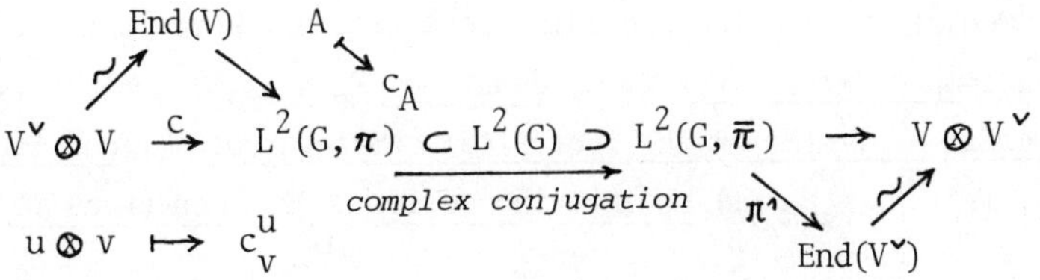

Schur's orthogonality relations (5.6)

$$(c^u_v \mid c^x_y) = \frac{1}{\dim\pi}\overline{(u \mid x)}\,(v \mid y). \qquad (\dim\pi = \dim V)$$

show that

$$c\ (u \otimes v \mapsto c^u_v)\ \underline{\text{is}}\ \ (\dim\pi)^{-\frac{1}{2}}\cdot\underline{\text{isometry map}}\ .$$

The inverse of c is nearly the extension π^1 of π. Let us compute $\pi^1(c^u_v)$. For this purpose, we apply this operator to a vector y and compute the scalar product of the result with a vector x

$$(x \mid \pi(c^u_v)\,y) = \int c^u_v(s)\,(x \mid \pi(s)y)\,ds = (\bar{c}^u_v \mid c^x_y)\ .$$

This quantity will vanish when $\bar{\pi}$ is not equivalent to π (5.6.a). It would have been more appropriate to compute $\pi(\bar{c}^u_v)$! Indeed, Schur's relations give then

$$(x \mid \pi(\bar{c}^u_v)y) = (c^u_v \mid c^x_y) = (\dim\pi)^{-1}\overline{(u \mid x)}\,(v \mid y) =$$

$$= (\dim\pi)^{-1}(x \mid (v \mid y)\,u) = (x \mid (\dim\pi)^{-1}v \otimes u\,(y))$$

which implies

$$\pi(\bar{c}^u_v) = (\dim\pi)^{-1}\ v \otimes u$$

and by linearity

$$\pi(\bar{c}_A) = (\dim\pi)^{-1}\ A^* \qquad .$$

Since the coefficients $\bar{c}_A$ are the coefficients of $\bar{\pi}$, we see that on $L^2(G,\bar{\pi})$, $f \mapsto \pi(f)$ is of the form

$$\pi^1\Big|_{L^2(G,\bar{\pi})} = (\dim\pi)^{-\frac{1}{2}}\cdot\text{isometry map}\ .$$

The composite

$$\begin{array}{ccccccc} \mathrm{End}(V) & \rightarrow & L^2(G,\pi) & \xrightarrow{conj.} & L^2(G,\bar{\pi}) & \rightarrow & \mathrm{End}(V^\vee) \\ A & \mapsto & c_A & & \bar{c}_A = f & \mapsto & \pi^1(f) \end{array}$$

can be identified with

$$(\dim\pi)^{-1}\cdot(A \mapsto A^*) = (\dim\pi)^{-1}\cdot\ \text{isometry map}\ .$$

The <u>Plancherel theorem</u> (and <u>Fourier inversion theorem</u>) follows immediately from the preceding computations. Here it is. <u>Let</u> G <u>be a compact group</u>. <u>For</u> $\pi \in \hat{G}$, <u>denote by</u> $P_\pi : L^2(G) \rightarrow L^2(G,\pi)$ <u>the orthogonal projector on the isotypical component of</u> π, <u>and for</u> $f \in L^2(G)$, <u>let</u> $f_\pi = P_\pi(f)$ <u>so that the series</u> $\sum_{\hat{G}} f_\pi$ <u>converges to</u> f in quadratic mean (i.e. in $L^2(G)$). Then

a) $f_\pi(x) = \dim\pi \cdot \mathrm{Tr}(\pi(\check{f})\pi(x)) \qquad (\check{f}(x) = f(x^{-1}))$,

b) $\|f_\pi\|_2^2 = \dim\pi \, \|\pi(\check{f})\|_2^2$ (Hilbert-Schmidt norm) ,

c) $\|f\|_2^2 = \sum_{\pi\in\hat{G}} \dim\pi \, \|\pi(f)\|_2^2$ (Parseval equality) .

(In particular, all components f_π are continuous functions cf.(4.6).)

Proof. The orthogonal projection f_π of f in $L^2(G,\pi)$, is the element c_A of this space having same scalar product with all elements c_B of $L^2(G,\pi)$. (Burnside's theorem shows that the endomorphism A is completely determined by the condition $c_A = f_\pi$ cf.(5.2).) Let us determine A as function of f. We must have

$$(c_B \mid f) = (c_B \mid f_\pi) = (c_B \mid c_A) = (\dim\pi)^{-1}\,\mathrm{Tr}(B^*A) .$$

But

$$\begin{aligned}(c_B \mid f) &= \int \overline{c_B(x)}\, f(x)\, dx = \int \overline{\mathrm{Tr}(B\pi(x))}\, f(x)\, dx = \\ &= \int \mathrm{Tr}(\pi(x^{-1})B^*) f(x)\, dx = \mathrm{Tr}(B^*\!\int \pi(x^{-1})\, f(x)\, dx\,) = \\ &= \mathrm{Tr}(B^*\, \pi(\check{f})\,) \quad .\end{aligned}$$

Comparing the two results obtained for all B, we indeed find

$$A = \dim\pi \cdot \pi(\check{f}) \quad .$$

This gives

$$f_\pi(x) = c_A(x) = \mathrm{Tr}(A\pi(x)) = \dim\pi \cdot \mathrm{Tr}(\pi(\check{f})\pi(x))$$

as asserted in a). Moreover, Schur's relations show that

$$\|f_\pi\|_2^2 = \|c_A\|_2^2 = (\dim\pi)^{-1}\|A\|_2^2 = \dim\pi \cdot \|\pi(\check{f})\|_2^2 \quad .$$

Thus b) is proved and c) follows from the observation that f and $\check{f}$ have the same quadratic norm (Ex. 5 of sec.1): we can permute f and $\check{f}$. Also observe that the dimensions of π and of $\check{\pi}$ are the same ! q.e.d.

Let us observe explicitly that this theorem does *not* imply that the series $\sum f_\pi$ of continuous functions converges for nearly all x in G (convergence in quadratic mean does not imply simple convergence outside a set of measure zero). Even when f is continuous, simple convergence of this series $\sum f_\pi$ outside a negligible set is a deep statement (proved when G is the circle group, cf. infra.).

EXAMPLE

Let now G be the circle group $\mathbb{R}/\mathbb{Z}$. As this group is commutative, all its irreducible representations have dimension 1 (dimension 1 representations are always *irreducible* and are also called *characters* : a more general definition of characters will be given in next sec.7). The dual $\hat{G}$ of this circle group is thus

$$\hat{G} = \{\pi_n : n \in \mathbb{Z}\} \quad \text{where} \quad \pi_n(x) = e^{2\pi inx} \quad .$$

The Fourier coefficients c_n of a continuous function f (or $f \in L^1(G)$) are given by the usual formulas

$$c_n = c_n(f) = \int_0^1 f(x)\, e^{-2\pi inx}\, dx =$$

$$= \int_0^1 e^{2\pi inx}\, f(-x)\, dx = \pi_n(\check{f}) \quad .$$

Thus the Fourier series of f is identical with the series $\sum f_\pi$ given in the Plancherel theorem

$$\sum c_n(f)\, e^{2\pi inx} = \sum \pi_n(\check{f})\, \pi_n(x) =$$

$$= \sum \dim \pi_n \cdot \mathrm{Tr}(\pi_n(\check{f})\pi_n(x)) \quad .$$

This series converges in quadratic mean, and it is only recently (1966) that Carleson has been able to prove convergence for nearly all x (cf. Fefferman 1974, p.105). This answered a question raised by Lusin in 1915. When f is continuous (and periodic of period 1), Fejer's theorem asserts that the Cesaro means converge uniformly to f. But for a compact group G, $\hat{G}$ has no natural ordering and Fejer's theorem has no generalization in this direction.

EXERCISES

1. Let G be a compact group, f and $g \in L^1(G)$. For any representation σ of G (in a Banach space), prove

$$\sigma(f * g) = \sigma(f) \cdot \sigma(g) .$$

If σ is unitary, prove also $\sigma(f^*) = \sigma(f)^*$ (recall that $f^*(x) = \overline{f(x^{-1})}$).

2. Show that the "extensions" of the regular representations of a compact group G are given by

$$\ell(f)(\varphi) = f * \varphi \quad , \quad \rho(g)(\varphi) = \varphi * \check{g}$$

where f and $g \in L^1(G)$ and $\varphi \in L^2(G)$ (recall that $\check{g}(x) = g(x^{-1})$). Conclude from this that

$$\| f * \varphi \|_2 \leqslant \| f \|_1 \| \varphi \|_2 .$$

Moreover, using 1), deduce the associativity

$$(f * g) * \varphi = f * (g * \varphi)$$

($f, g \in L^1(G)$, $\varphi \in L^2(G)$ or all three functions in $C(G)$!). Also check that for any representation σ of G

$$\sigma(\ell(x) f) = \sigma(x)\sigma(f) \quad , \quad \sigma(\rho(x) f) = \sigma(f)\sigma(x^{-1}) .$$

3. Let G be a compact group and denote by $L^1_{inv} = L^1_{inv}(G)$ the closure in $L^1(G)$ of the subspace of continuous functions f satisfying $f(xy) = f(yx)$ (or equivalently, $f(y^{-1}xy) = f(x)$) for all x and y in G. Show that L^1_{inv} is contained in the center of $L^1(G)$ (as convolution algebra : prove $f * g = g * f$ for $f \in L^1_{inv}$ and $g \in L^1(G)$). For any irreducible representation $\pi : G \longrightarrow Gl(V)$ prove that

$$\pi(f) = (\dim \pi)^{-1} \langle \chi, f \rangle \, 1_V \qquad (f \in L^1_{inv})$$

where $\chi(g) = \mathrm{Tr}\, \pi(g)$ and $\langle \chi, f \rangle = \int \chi(g) f(g)\, dg$.
(<u>Hint:</u> use Schur's lemma to prove that $\pi(f)$ is a scalar operator and then take traces to determine the value of the constant in this multiple of the identity.)

4. Let $\sigma : G \longrightarrow Gl(V)$ be a unitary representation of a compact group G. Check that $\sigma(1) = P$ (1 = constant function 1 in $L^1(G)$) is the orthogonal projector $V \longrightarrow V^G$ on the subspace of G-invariants of V. (<u>Hint</u>: show that $1 * 1 = 1$ and $1^* = 1$ in $L^1(G)$; more generally, $1 * f = f * 1$ is the constant function $\int f(x)\,dx$.)

5. Show that the "extended" left regular representation

$$\ell = \ell^1 \ : \ L^1(G) \longrightarrow End(L^2(G))$$

has trivial kernel $\{0\}$ (<u>Hint</u>: Let $0 \neq f \in L^1(G)$ and construct a sequence $(g_n) \subset C(G)$ with $g_n \geqslant 0$, $\int g_n(x)\,dx = 1$, and $\ell(f)(g_n) = f * g_n \longrightarrow f \neq 0$.) Conclude that if $0 \neq f \in L^1(G)$, there exists a π in $\hat{G}$ such that $\pi(f) \neq 0$ (use the decomposition of the regular representation given in sec.5). Finally, prove that

$$L^1(G) \text{ commutative } \Longleftrightarrow G \text{ commutative} \ .$$

6. Let G be a compact group, $\pi \in \hat{G}$ and consider the adjoint representation of G in $End(V)$ ($V = V_\pi$) defined by the following composition

$$\begin{array}{llllll} Ad : & G & \longrightarrow & G \times G & \longrightarrow & End(V) \\ & s & \longmapsto & (s,s) & \longmapsto & (A \longmapsto \pi(s)\,A\,\pi(s)^{-1}) \\ & & & (s,t) & \longmapsto & (A \longmapsto \pi(t)\,A\,\pi(s)^{-1}) \end{array} \ .$$

Prove that the multiplicity of the identity representation in this adjoint representation is 1 (this identity representation acts on the subspace of scalar operators: Schur's lemma).

7. The decomposition of the biregular representation of a compact group G in $L^2(G)$ gives the decompositions of the left (resp. right) regular representation simply by composition with

$$\begin{array}{llll} i_1 : G \longrightarrow G \times G & & (\text{resp. } i_2 : G \longrightarrow G \times G & \\ \quad s \longmapsto (s,e) & & \quad s \longmapsto (e,s) &) \ . \end{array}$$

Conclude that

$$\ell \cong \widehat{\bigoplus}\, \check{\pi} \otimes 1 \cong \widehat{\bigoplus} \dim \pi \cdot \check{\pi} = \widehat{\bigoplus} \dim \pi \cdot \pi \ ,$$

$$\rho \cong \widehat{\bigoplus}\, 1 \otimes \pi \cong \widehat{\bigoplus} \dim \pi \cdot \pi \ .$$

(Compare with Ex.1 of sec.2 .)

7 CHARACTERS AND GROUP ALGEBRAS

Let (π,V) be a finite dimensional representation of a compact group G. The character $\chi = \chi_\pi$ of π is the (complex valued) continuous function on G defined by

$$\chi(x) = \mathrm{Tr}(\pi(x)) \quad .$$

(This is the function c_A for A = id. $\in$ End(V), cf. sec.6.) When dim(V) = 1, χ and π can be identified: in this case, χ is a homomorphism.

Quite generally, since the trace satisfies the identity Tr(AB) = Tr(BA), we see that the characters of two equivalent representations are equal. Moreover, characters satisfy

$$\chi(xy) = \chi(yx) \quad \text{or} \quad \chi(y^{-1}xy) = \chi(x) \qquad (x, y \in G).$$

Thus characters are *invariant* functions

$$\chi \in C_{inv} = \{f \in C(G) \ : \ f(y^{-1}xy) = f(x)\ ,\ x \text{ and } y \text{ in } G\}.$$

We shall also have to use

$$L^1_{inv} = \text{closure of } C_{inv} \text{ in } L^1(G) \quad ,$$

$$L^2_{inv} = \text{closure of } C_{inv} \text{ in } L^2(G)$$

(cf. Ex.3 of sec.6). Invariant functions are also called *central* functions (they belong to the center of $L^1(G)$ with respect to convolution).

Still quite generally, the character of the contragredient $\check{\pi}$ of π is given by

$$\chi_{\check{\pi}}(x) = \mathrm{Tr}\,\check{\pi}(x) = \mathrm{Tr}\ {}^t\pi(x^{-1}) = \mathrm{Tr}\,\pi(x^{-1}) = \chi(x^{-1})$$

hence $\chi_{\check{\pi}} = \check{\chi}$. When π is unitary, $\pi(x^{-1}) = \pi(x)^*$ ($\check{\pi}$ is equivalent to $\bar{\pi}$) and $\check{\chi}$ is the complex conjugate of χ. One can also check without difficulty that for two finite dimensional representations π, σ of G

$$\chi_{\pi\oplus\sigma} = \chi_\pi + \chi_\sigma \ , \quad \chi_{\pi\otimes\sigma} = \chi_\pi \cdot \chi_\sigma \quad .$$

When π is irreducible, Schur's lemma shows that elements z in the center Z of G are mapped on scalar operators by π : $\pi(z) = \lambda_z 1_V$ so that

$\chi(z) = \lambda_z \cdot \dim V$. Thus the restriction of $(\dim V)^{-1}\chi$ to the center Z is a homomorphism

$$\lambda : Z \to \mathbb{C}^\times .$$

This is the <u>central character</u> of π : it is independent from the special model chosen in the equivalence class of π. In particular if $\lambda(Z)$ is *not contained in* $\{\pm 1\}$, π and $\bar{\pi}$ are *not equivalent* (their central characters are different). Also observe that $\chi(e) = \dim V$ (= $\dim \pi$).

(7.1) <u>Proposition</u>. <u>Any continuous central function</u> $f \in C_{inv}$ <u>on a compact group</u> G <u>is a uniform limit of linear combinations of characters of irreducible representations of</u> G.

<u>Proof</u>. Let $\varepsilon > 0$. By the Peter-Weyl theorem (4.2), we know that there is a finite dimensional representation (σ, V) and a $A \in \mathrm{End}(V)$ with

$$|f(x) - \mathrm{Tr}(A\sigma(x))| < \varepsilon \qquad (x \in G) .$$

In this expression, replace x by one of its conjugates yxy^{-1} :

$$|f(x) - \mathrm{Tr}(A\sigma(yxy^{-1}))| = |f(x) - \mathrm{Tr}(\sigma(y^{-1})A\sigma(y)\sigma(x)| < \varepsilon .$$

Integrating over y, we get

$$|f(x) - \mathrm{Tr}(B\sigma(x))| < \varepsilon \quad \text{where} \quad B = \int \sigma(y^{-1})\, A\, \sigma(y)\, dy .$$

By invariance of the Haar measure, the operator B commutes to all operators $\sigma(x)$. Hence, if we decompose σ into irreducible components (or rather, into isotypical components)

$$\sigma \cong \bigoplus n_\pi \pi \cong \bigoplus \pi \otimes 1_{n_\pi} ,$$

the operator B will have the form

$$B = \bigoplus 1_{\dim\pi} \otimes B_\pi \qquad \text{(cf. Ex.1 of this sec.)}$$

and

$$B\sigma(x) = \sigma(x)B \cong \bigoplus \pi \otimes B_\pi ,$$

$$\mathrm{Tr}(B\sigma(x)) = \sum a_\pi \chi_\pi(x) \qquad (a_\pi = \mathrm{Tr}\, B_\pi) .$$

This shows

$$|f(x) - \sum_{\text{finite}} a_\pi \chi_\pi(x)| < \varepsilon .$$

(7.2) <u>Theorem</u>. <u>Let</u> π <u>and</u> σ <u>be two finite dimensional representations of a compact group</u> G <u>with respective characters</u> χ_π <u>and</u> χ_σ. <u>Then</u>

$$(\chi_\pi | \chi_\sigma) = \dim \mathrm{Hom}_G(V_\pi, V_\sigma) .$$

Proof. From (5.4) (or alternatively, Ex.2.c) of sec.5), we know that the integral

$$\int \check{\pi}(x) \otimes \sigma(x)\, dx$$

is an expression for the projector

$$\begin{array}{lll} \natural : & \check{V_\pi} \otimes V_\sigma & \longrightarrow (\check{V_\pi} \otimes V_\sigma)^G \\ & \mathrm{Hom}(V_\pi, V_\sigma) & \longrightarrow \mathrm{Hom}_G(V_\pi, V_\sigma) \end{array} .$$

The dimension of the image space is the trace of this projector. Thus

$$(\chi_\pi \mid \chi_\sigma) = \int \overline{\chi_\pi(x)}\, \chi_\sigma(x)\, dx = \dim \mathrm{Hom}_G(V_\pi, V_\sigma) . \quad \text{q.e.d.}$$

As in the case of finite groups, there are many important corollaries of this theorem. Their proof is immediate.

(7.3) Corollary 1. Let π be a finite dimensional representation of G. Then

$$\pi \text{ is irreducible} \iff \|\chi_\pi\|_2 = (\chi_\pi \mid \chi_\pi)^{\frac{1}{2}} = 1 .$$

(7.4) Corollary 2. Let $\pi, \sigma \in \hat{G}$. Then

$$(\chi_\pi \mid \chi_\sigma) = \delta_{\pi\sigma} \quad (= 1 \text{ if } \pi \text{ equivalent to } \sigma,\ 0 \text{ otherwise}).$$

(7.5) Corollary 3. Let σ be a finite dimensional representation of G and $\sigma = \oplus n_\pi \pi$ (summation over a finite subset $I \subset \hat{G}$) be a decomposition into irreducible components. Then

a) $n_\pi = (\chi_\sigma \mid \chi_\pi)$ is well determined ,

b) $\|\chi_\sigma\|^2 = \sum n_\pi^2$ (summation over I) .

Proof. Let $V_\sigma = \oplus (V_\tau \otimes \mathbb{C}^{n_\tau})$. Since each G-morphism $V_\sigma \longrightarrow V_\pi$ must vanish on all isotypical components $V_\tau \otimes \mathbb{C}^{n_\tau}$ where τ is not equivalent to π, we have

$$\begin{aligned} \mathrm{Hom}_G(V_\sigma, V_\pi) &= \mathrm{Hom}_G(V_\pi \otimes \mathbb{C}^{n_\pi}, V_\pi) = \\ &= \mathbb{C}^{n_\pi} \otimes \mathrm{Hom}_G(V_\pi, V_\pi) = \mathbb{C}^{n_\pi} . \end{aligned}$$

Assertion a) results from this. Moreover, b) follows from Pythagoras theorem (and (7.4) above !).

(7.6) Corollary 4. The set of characters $(\chi_\pi)_{\pi \in \hat{G}}$ is an orthonormal basis of the Hilbert space L^2_{inv} (closure of C_{inv} in $L^2(G)$): every invariant square summable function f on G can be expanded in a series

$$f = \sum_{\hat{G}} (\chi_\pi | f)\, \chi_\pi \qquad \text{(convergence in } L^2(G)\text{)} .$$

(7.7) Theorem. Let G be a compact group. For $\pi \in \hat{G}$, let P_π denote the projector $L^2(G) \to L^2(G,\pi)$ onto the isotypical component of π (in the right regular representation). Then P_π is given by convolution with the normalized character $\vartheta_\pi = \dim \pi \cdot \chi_\pi$:

$$P_\pi : f \mapsto f_\pi = P_\pi f = f * \vartheta_\pi .$$

Proof. We have already seen (part a) of Plancherel's theorem, sec.6) that $f_\pi(x) = \dim \pi \cdot \mathrm{Tr}(\pi(\check{f})\pi(x))$.

Thus,

$$f_\pi(x) = \dim \pi\, \mathrm{Tr} \int f(y^{-1})\pi(y)\pi(x)\, dy =$$

$$= \dim \pi \int f(y)\, \mathrm{Tr}\, \pi(y^{-1}x)\, dy = f * \vartheta_\pi(x) \qquad \text{q.e.d.}$$

Still denoting by ρ the right regular representation, we can write $f * \vartheta_\pi = \rho(\check{\vartheta}_\pi)(f)$. This also shows that

$$P_\pi : L^2(G) \longrightarrow L^2(G,\pi)$$

is also given by

$$P_\pi = \rho(e_\pi) \quad \text{where} \quad e_\pi = \dim \pi \cdot \check{\chi}_\pi = \dim \pi \cdot \overline{\chi_\pi}.$$

Since ϑ_π is a linear combination of coefficients of π, hence belongs to $L^2(G,\pi)$, we have

$$\vartheta_\pi = P_\pi(\vartheta_\pi) = \vartheta_\pi * \vartheta_\pi .$$

By conjugation, we also have $e_\pi * e_\pi = e_\pi$. Moreover, if π and $\sigma \in \hat{G}$ are not equivalent, $\vartheta_\pi * \vartheta_\sigma = 0$ and $e_\pi * e_\sigma = 0$.

Let us also observe that if we had worked with the *left* regular representation, the isotypical component of type π in this representation would have been $L^2(G,\bar{\pi})$ and the projector onto this component could have been written

$$Q_\pi(f) = f_{\bar{\pi}} = f * \vartheta_{\bar{\pi}} = \vartheta_{\bar{\pi}} * f = \ell(\vartheta_{\bar{\pi}})\, f .$$

Hence

$$Q_\pi = \ell(e_\pi) \quad \text{is the projector } L^2(G) \to L^2(G,\bar{\pi}) .$$

More generally, we have the following decomposition theorem.

(7.8) Theorem. Let (σ, V) be a representation of a compact group G in a Banach space (or more generally in a *quasi-complete locally convex* space). Then

1) $V_\pi = \sigma(e_\pi)V$ is a closed subspace $(\pi \in \hat{G})$,
2) $\sum V_\pi = \oplus V_\pi$ is a direct sum in V ,
3) $\oplus V_\pi$ is dense in V ,
4) $v \mapsto (v_\pi) = (\sigma(e_\pi)v) : V \to \prod V_\pi$ is injective .

Moreover, any $a \in V_\pi$ generates a G-invariant subspace of dimension $\leq (\dim \pi)^2$. When σ is unitary (V being a Hilbert space), V is the Hilbert sum $\hat{\oplus} V_\pi$ (completion of the orthogonal direct sum $\oplus V_\pi$) : in this case, we thus have $v = \sum v_\pi$ with a convergence of the series in V.

Remark. Taking for G the circle group and for V the Banach space C(G) of continuous functions (with the uniform norm), we see that the series $\sum v_\pi$ cannot be expected to converge in V = C(G) (Example at the end of sec.6). The series is thus only considered in the Hilbert space case.

Proof. Projectors in a vector space are characterized among linear operators by the identity $P^2 = P$. For such an operator, the image of P is the kernel of I - P, hence is closed if P is continuous. Point 1) of the theorem results from this observation with

$$P = \sigma(e_\pi) = \sigma(e_\pi * e_\pi) = \sigma(e_\pi)\cdot\sigma(e_\pi) = P^2 .$$

To prove 2), we observe that in a finite sum $\sum v_\tau = 0$ (sum over a finite subset I of $\hat{G}$) of elements $v_\tau \in V_\tau$, we can write $v_\tau = \sigma(e_\tau)v_\tau$ hence

$$0 = \sigma(e_\pi)\sum \sigma(e_\tau)v_\tau = \sum \sigma(e_\pi * e_\tau)v_\tau = v_\pi .$$

Thus the subspaces V_π are linearly independent and their sum is direct. The other points will result from the fact that $L^2(G,\pi)$ is irreducible under the *biregular* representation $\ell \times \rho$ of $G \times G$, hence is generated by the left *and* right translates of any of its non-zero elements. For example, the left translates of χ_π already generate $L^2(G,\pi)$ (since $\chi_\pi(st) = \chi_\pi(ts)$, there is no distinction between left and right translates of χ_π). Then, for $f \in C(G)$, $u \in V^\vee$ and $v \in V$ we have

$$\int \overline{f(x)}\langle u, \sigma(x)v\rangle\, dx = \langle u, \int \overline{f(x)}\, \sigma(x)v\, dx\rangle$$

hence

$$(f \mid c_v^u) = \langle u, \sigma(\bar{f})\, v \rangle \quad .$$

To prove 3), we apply the preceding equality to $f = \rho(s^{-1})\chi_\pi$:

$$(\rho(s^{-1})\chi_\pi \mid c_v^u) = \langle u, \ \sigma(\rho(s^{-1})\overline{\chi_\pi})\, v \rangle =$$

$$= \langle u, \ \sigma(\overline{\chi_\pi})\, \sigma(s)\, v \rangle \ .$$

Thus if the linear form u vanishes on $\oplus V_\pi$, we have

$$c_v^u \perp L^2(G, \pi) \qquad (\pi \in \hat{G})$$

$$c_v^u = 0 \ , \text{ in particular } \langle u,v \rangle = c_v^u(e) = 0 \qquad (v \in V)$$

whence $u = 0$. Density of $\oplus V_\pi$ results then from the Hahn-Banach theorem. To prove 4), we consider $f = \ell(s)\chi_\pi \ (= \rho(s^{-1})\chi_\pi)$:

$$(\ell(s)\chi_\pi \mid c_v^u) = \langle u, \sigma(\ell(s)\overline{\chi_\pi})\, v \rangle = \langle u, \ \sigma(s)\sigma(\overline{\chi_\pi})\, v \rangle .$$

Hence if $v \in V$ is such that all components $v_\pi = 0$ (for all $\pi \in \hat{G}$) , we have

$$c_v^u \perp \ell(s)\chi_\pi \quad , \quad c_v^u \perp L^2(G, \pi) \quad (\text{all } \ \pi \in \hat{G}) ,$$

hence

$$c_v^u = 0 \ , \text{ in particular } \ \langle u,v \rangle = c_v^u(e) = 0$$

(for every continuous linear form $u \in E'$ on E). Again $v = 0$ by the Hahn-Banach theorem. Finally, take $a \in V_\pi$. The relation $a = \sigma(e_\pi)\, a$ can be written

$$a = \int \sigma(x)\, e_\pi(x)\, a\, dx \quad ,$$

hence

$$\sigma(y)\, a = \int e_\pi(x)\, \sigma(yx)\, a\, dx = \int e_\pi(y^{-1}x)\, \sigma(x)\, a\, dx \quad .$$

As the translates $\ell(y)e_\pi$ generate the subspace $L^2(G,\bar{\pi})$ of dimension equal to $(\dim \pi)^2$, we see that the $\sigma(y)\, a$ remain in a subspace of dimension $\leqslant (\dim \pi)^2$. More precisely,

$$f \longmapsto \int f(x)\, \sigma(x)\, a\, dx = \sigma(f)\, a$$

is a G-morphism

$$L^2(G, \bar{\pi}) \longrightarrow V_\pi \quad : \quad \pi \otimes 1 = \ell\big|_{L^2(G, \bar{\pi})} \longrightarrow \sigma\big|_{V_\pi} \quad .$$

The image $V_\pi(a)$ of this G-morphism is thus a sum of at most $\dim \pi$

representations equivalent to π and $\sigma|_{V_\pi}$ is isotypical of type π (as sum of the isotypical representations acting in all $V_\pi(a)$, $a \in V_\pi$). This concludes the proof of theorem (7.8).

(7.9) Corollary. Any irreducible representation of a compact group G in a quasi-complete locally convex vector space is finite dimensional.

Proof. Let V be the space of the irreducible representation σ and take $0 \neq a \in V$. There must exist a $\pi \in \hat{G}$ (hence π finite dimensional) with $a_\pi \neq 0$ (part 4 of the theorem (7.8)). The corresponding invariant subspace V_π is not $\{0\}$, hence $V = V_\pi = V_\pi(a)$ by irreducibility. But we have seen that $\dim V_\pi(a) \leqslant (\dim \pi)^2$ is finite.

AN APPLICATION: STONE'S THEOREM

We take for G the circle group $\mathbb{R}/\mathbb{Z}$, and consider a unitary representation U of this group in a Hilbert space H :

$$U : \mathbb{R} \longrightarrow Gl(H) \quad , \quad t \longmapsto U_t$$

satisfies

$$U_{t+1} = U_t \quad , \quad U_{t+s} = U_t U_s \qquad (s,t \in \mathbb{R}) .$$

The structure of this *one parameter periodic* subgroup $t \longmapsto U_t$ is as follows.

We already know that the dual of the circle group is

$$\hat{G} = \{\pi_n : n \in \mathbb{Z}\} \quad , \quad \pi_n(t) = e^{2\pi int} .$$

Hence $e_n = \dim \pi_n \operatorname{Tr}(\bar{\pi}_n)$ is given by $e_n(t) = e^{-2\pi int}$ and $P_n = U(e_n)$ is the orthogonal projector on the isotypical component of type π_n. For $v \in H$, we have thus

$$v = \sum v_n = \sum P_n(v) \quad \text{and} \quad U_t(v) = \sum P_n(v)\, e^{2\pi int}$$

with

$$v_n = P_n(v) = \int e^{-2\pi int}\, U_t(v)\, dt$$

(the integral being computed over $\mathbb{R}/\mathbb{Z}$ or simply over the interval $[0,1]$). Thus we can write

$$U_t = \sum e^{2\pi int} P_n$$

with a simple convergence ($\sum e^{2\pi int} P_n(v)$ converges in V for every $v \in V$).

This result can be stated in a slightly different way. Put

$$E_\lambda = \sum_{2\pi n \leqslant \lambda} P_n \ .$$

Then we can write

$$U_t = \int e^{i\lambda t}\, dE_\lambda = e^{iAt}$$

with a certain self-adjoint (unbounded) operator $A = \int \lambda \, dE_\lambda$. The non-periodic case of a one parameter unitary group $t \mapsto V_t$ is easily reduced to the periodic case by the definition

$$U_t = V_t V_1^{-t} \ .$$

The definition of the power V_1^{-t} requires some spectral theory for the unitary operator V_1 : for these classical notations, and for the treatment of one parameter groups of unitary transformations, including the reduction to the periodic case, one can consult Riesz-Nagy 1975, pp. 375-380 .

Characters of irreducible representations of compact groups can also be characterized by an integral equation.

(7.10) *Theorem* (*H. Weyl*). *Let* χ *be the character of an irreducible representation of a compact group* G. *Then*

$$\chi(s)\,\chi(t) = \chi(e) \int_G \chi(xsx^{-1}t)\, dx \ .$$

Conversely, let $\varphi \neq 0$ *be an integrable function satisfying*

$$\varphi(s)\,\varphi(t) = \int_G \varphi(xsx^{-1}t)\, dx \ .$$

Then there is a (*unique*) π *in* $\hat{G}$ *with* $\varphi = \chi_\pi / \dim \pi$.

Proof. We have seen (5.5) that for $\pi \in \hat{G}$ and $A \in \mathrm{End}(V_\pi)$,

$$A^{\natural} = \int \pi(x)\, A\, \pi(x)^{-1}\, dx = (\dim \pi)^{-1}\, \mathrm{Tr}(A)\, \mathrm{id}_V$$

is a scalar operator in $V = V_\pi$. Consequently, for any $B \in \mathrm{End}(V)$

$$\int \pi(x)\, A\, \pi(x)^{-1}\, B\, dx = (\dim \pi)^{-1}\, \mathrm{Tr}(A)\, B \ .$$

Take $A = \pi(s)$ and $B = \pi(t)$:

$$\int \pi(xsx^{-1}t)\, dx = (\dim \pi)^{-1} \chi(s)\, \pi(t)$$

whence the result by taking traces ($\dim \pi = \chi(e)$) . Conversely, suppose $0 \neq \varphi \in L^1(G)$, and φ satisfies the integral equation. There exists a $\pi \in \hat{G}$ with $\pi(\varphi) \neq 0$ (cf. Ex. 5 of sec.6) . With this π

we can write

$$\varphi(s)\,\pi(\varphi) = \int \varphi(s)\varphi(t)\pi(t)\,dt = \int dt\ \pi(t)\int dx\,\varphi(xsx^{-1}t) =$$
$$= \int dx \int dt\,\pi(t)\,\varphi(xsx^{-1}t) = \int dx \int dt\ \pi((xsx^{-1})^{-1}t)\,\varphi(t) =$$
$$= \int dx \int dt\ \varphi(t)\,\pi(xs^{-1}x^{-1}t) = \int dx\ \pi(x)\pi(s^{-1})\pi(x)^{-1}\pi(\varphi)\ .$$

But now

$$\int dx\ \pi(x)\pi(s^{-1})\,\pi(x)^{-1} = (\dim\pi)^{-1}\ \mathrm{Tr}\ \pi(s^{-1}) = \frac{\overline{\chi(s)}}{\chi(e)}\ .$$

Hence

$$\varphi(s)\,\pi(\varphi) = \chi(e)^{-1}\ \overline{\chi(s)}\ \pi(\varphi) \in \mathrm{End}(V)\ .$$

This proves that $\chi(e)\ \varphi(s) = \overline{\chi(s)}$ and $\varphi = \chi_{\check{\pi}}/\chi_{\check{\pi}}(e)$. q.e.d.

It is also possible to express the Plancherel theorem (sec.6) with characters of irreducible representations. Let us recall that the Dirac measure (or distribution) on a compact group G is given by the linear form

$$\delta\ :\ C(G) \longrightarrow \mathbb{C}\ ,\ f \longmapsto f(e)$$

of evaluation at the neutral element. For $f \in L^2(G)$, we have seen that

$$\|f\|^2 = \sum\ \dim\pi\cdot\|\pi(f)\|_2^2\ .$$

By *polarization* (replace f by $f \pm g$ and by $f \pm i\,g$ and add the results), we also have

$$(f\mid g) = \sum\ \dim\pi\cdot\mathrm{Tr}(\pi(f)^*\,\pi(g))\ .$$

But $(f\mid g) = f^* * g\,(e) = \varphi(e)$ (with $f^*(x) = \overline{f(x^{-1})}$ and $\varphi = f^* * g$ continuous on G), hence

$$\pi(f)^*\pi(g) = \pi(f^*)\pi(g) = \pi(f^* * g) = \pi(\varphi),$$

$$\varphi(e) = \sum\ \dim\pi\ \mathrm{Tr}\ \pi(\varphi)$$

for all continuous functions φ of the special form $f^* * g$ with some f and $g \in L^2(G)$. Moreover, the sum giving $\varphi(e)$ converges absolutely (since the sum giving $(f\mid g)$ converges absolutely). Writing

$$\mathrm{Tr}\ \pi(\varphi) = \mathrm{Tr}\int \pi(x)\varphi(x)\,dx = \int \chi(x)\varphi(x)\,dx = \langle \chi, \varphi\rangle$$

we see that

$$\langle \delta, \varphi\rangle = \varphi(e) = \sum\ \dim\pi\ \langle \chi_\pi, \varphi\rangle \qquad \text{(sum over } \hat{G})\ .$$

In a certain sense, we can thus write

$$\delta = \sum \dim\pi \cdot \chi_\pi \qquad \text{(sum over } \hat{G}\text{)}.$$

Replace the special function φ by one of its right translates (if $\varphi = f^* * g$, this amounts to replacing g by the corresponding right translate: $\rho(s)\varphi = f^* * \rho(s)g$)

$$\varphi(s) = \rho(s)\varphi(e) = \sum \dim\pi \operatorname{Tr}\pi(\rho(s)\varphi) =$$

$$= \sum \dim\pi \operatorname{Tr}(\pi(\varphi)\pi(s^{-1})) \qquad \text{(by ex. 2 of sec.6)} .$$

Replace now φ by $\check{\varphi}$ and s by s^{-1}

$$\varphi(s) = \check{\varphi}(s^{-1}) = \sum \dim\pi \operatorname{Tr}(\pi(\check{\varphi})\pi(s)) .$$

We already knew that this formula holds (part a) of the Plancherel theorem, sec.6), but we have now proved an absolute convergence at each point when φ is a finite linear combination of continuous functions of the special form $f^* * g$ (f and g in $L^2(G)$). (Observe that left translations would have led to the same result since $\varphi(s) = \ell(s^{-1})\varphi(e)$, $\pi(\ell(s^{-1})\varphi) = \pi(s^{-1})\pi(\varphi)$ has same trace as $\pi(\varphi)\pi(s^{-1})$.)

When G is the circle group $\mathbb{R}/\mathbb{Z}$, the special functions $f^* * g$ are those which have an absolutely summable sequence of Fourier coefficients. It is not possible to characterize these functions as simply in the general case (i.e. when G is any compact group).

ALGEBRAS ATTACHED TO A COMPACT GROUP

Since several algebras connected with a compact group have played a rôle in our theory, we review them and introduce another universal construction.

The first group algebra of G is the convolution algebra $L^1(G)$. In a certain sense, it plays a rôle completely analogous to the group algebra of a finite group. We have seen that if σ is a unitary representation of G, and $f \in L^1(G)$, the operator $\sigma(f)$ has a uniform norm bounded by the L^1-norm of f

$$\|\sigma(f)\| \leqslant \|f\|_1 .$$

Let us define a new norm on $L^1(G)$ by

$$\|f\|_* = \operatorname{Sup} \|\sigma(f)\|$$

(the supremum being taken over all unitary representations σ of G). Thus we have $\|f\|_* \leq \|f\|_1$. We also denote by C*(G) the Banach algebra obtained by completion of the convolution algebra $L^1(G)$ with this new norm (the elements of C*(G) cannot be represented by functions over G). By definition, any unitary representation of G can be extended canonically to a representation of C*(G) (extending the corresponding representation of $L^1(G)$) . This algebra C*(G) is a *stellar algebra* : the involution $f \mapsto f^*$ ($f^*(x) = \overline{f(x^{-1})}$) extends to an involution $a \mapsto a^*$ of C*(G) and

$$\|a^*a\|_* = \|a\|_*^2 \qquad (a \in C^*(G)).$$

We shall now give a model (faithful representation) of the algebra C*(G) by means of operators in $L^2(G) = H$. For this purpose, we extend the left regular representation and obtain

$$\ell^* : C^*(G) \longrightarrow \mathrm{End}(H) \qquad (H = L^2(G)).$$

I claim that ℓ^* is an isomorphism onto a uniformly closed subalgebra of operators of the Hilbert space H (the involution of End(H) being the operation of taking the adjoint of an operator). By the decomposition theorem (7.8), we have

$$\|f\|_* = \sup_{\sigma \text{ unitary}} \|\sigma(f)\| = \sup_{\pi \in \hat{G}} \|\pi(f)\| .$$

Since the left regular representation ℓ has a decomposition in which all $\pi \in \hat{G}$ have a positive multiplicity (sec.5), we also have

$$\sup_{\pi \in \hat{G}} \|\pi(f)\| = \|\ell(f)\| = \|f * \ldots\| .$$

Hence

$$\|f\|_* = \|\ell(f)\| \qquad (f \in L^1(G))$$

and C*(G) can be identified to the closure (uniform norm) of the operator algebra $\ell(L^1(G))$ acting in the Hilbert space $H = L^2(G)$.

(7.11) Proposition. The image of the stellar algebra C*(G) by the left regular representation of a compact group G consists of the decomposable operators

$$\hat{\bigoplus} (A_\pi \otimes 1) \quad \text{of the Hilbert sum } \hat{\bigoplus}(V_\pi \otimes \check{V_\pi}) = \hat{\bigoplus} L^2(G, \overline{\pi})$$

with $A_\pi \in \mathrm{End}(V_\pi)$ and $\|A_\pi\| \rightarrow 0$ (for $\pi \rightarrow \infty$ in the discrete space $\hat{G}$).

<u>Thus the stellar algebra</u> C*(G) <u>can be identified with</u>

$$\widehat{\bigoplus_{\hat{G}}} \mathrm{End}(V_\pi) = \text{completion of } \bigoplus \mathrm{End}(V_\pi) \text{ for the uniform norm.}$$

<u>Proof</u>. For $f \in L^2(G)$ we have $\sum \dim\pi \, \|\pi(f)\|_2^2 = \|f\|^2 < \infty$, whence

$$\|\pi(f)\| \leqslant \|\pi(f)\|_2 \longrightarrow 0 \quad \text{for} \quad \pi \longrightarrow \infty \quad \text{in } \hat{G} \text{ (discrete).}$$

Consequently we must also have

$$\|\pi(f)\| \longrightarrow 0 \quad \text{for} \quad \pi \longrightarrow \infty \qquad \text{when } f \in L^1(G) \ ,$$

$$\|\pi(a)\| \longrightarrow 0 \quad \text{for} \quad \pi \longrightarrow \infty \qquad \text{when } a \in C^*(G) \ .$$

Thus $\ell^*(C^*(G))$ is contained in the indicated algebra of decomposable operators. Conversely, it is enough to check that the image of ℓ^* contains the algebraic sum $\bigoplus \mathrm{End}(V_\pi)$. But the character of a $\pi \in \hat{G}$ gives rise to an idempotent $e_\pi = \dim\pi \cdot \bar{\chi}_\pi \in C(G)$ for which

$$\ell(e_\pi) = \widehat{\bigoplus_\tau} (A_\tau \otimes 1)$$

with

$$A_\tau = \begin{cases} \mathrm{id} \in \mathrm{End}(V_\pi) & \text{if } \tau \text{ equivalent to } \pi \ , \\ 0 \in \mathrm{End}(V_\tau) & \text{otherwise} \end{cases} \ .$$

Taking finite linear combinations of translates of the e_π, we get all decomposable operators belonging to the algebraic sum $\bigoplus \mathrm{End}(V_\pi)$. q.e.d.

One can also consider the closure of $\ell(L^1(G))$ for the strong topology on End(H), $H = L^2(G)$ (a sequence $A_n \longrightarrow A$ for the strong topology when $A_n(x) \longrightarrow A(x)$ in H for all $x \in H$). This strong closure contains the identity operator $1 \in \mathrm{End}(H)$ (indeed, the series $\sum P_\pi$ converges strongly to 1). We obtain thus the *von Neumann algebra* generated by the left regular representation.

$$\mathcal{U}(G) = (\ell(G))'' = (\ell(L^1(G)))'' = (\ell^*(C^*(G)))''$$

(if $\mathfrak{A}$ is any set of operators in a Hilbert space H, we denote by $\mathfrak{A}'$ the set of operators $\mathfrak{A}' \subset \mathrm{End}(H)$ which commute with all elements of $\mathfrak{A}$, and $\mathfrak{A}'' = (\mathfrak{A}')'$; if $\mathfrak{A}$ is stable under $A \longmapsto A^*$, $\mathfrak{A}''$ is the von Neumann algebra generated by $\mathfrak{A}$). This strong closure can be identified to the algebra of decomposable operators

$$\widehat{\bigoplus}(A_\pi \otimes 1) \quad \text{with} \quad A_\pi \in \mathrm{End}(V_\pi) \text{ and } \sup_{\pi \in \hat{G}} \|A_\pi\| < \infty \ ,$$

and is thus the "product" of the von Neumann algebras $\mathrm{End}(V_\pi)$ $(\pi \in \hat{G})$. The use of the right regular representation ρ instead of ℓ would lead to symmetrical results. For example

$$\mathcal{V}(G) = \rho(G)'' = \mathcal{U}(G)' \cong \prod \mathbb{C} \otimes \mathrm{End}(V_\pi) .$$

The center of both $\mathcal{U}(G)$ ($= \mathcal{V}(G)'$) and $\mathcal{V}(G)$ ($= \mathcal{U}(G)'$) is

$$\mathfrak{Z} = \mathcal{U}(G) \cap \mathcal{V}(G) \cong \prod \mathbb{C} \otimes \mathbb{C} \cong \ell^\infty(\hat{G}) .$$

All these "products" of von Neumann algebras are characterized in the Cartesian product by $\mathrm{Sup} \|A_\pi\| < \infty$.

In the following <u>table</u>, we list all <u>functional algebras attached to a compact group</u>. The smallest one is the algebra A_G consisting of linear combinations of coefficients of finite dimensional representations (with the usual product: pointwise multiplication of values). On the vector space A_G , one also considers the *co-product*

$$A_G \rightarrow A_G \otimes A_G \quad : \quad f \mapsto \sum f_i \otimes g_i$$

where $f(st) = \sum f_i(s) g_i(t)$. With this co-product, A_G is a *Hopf algebra*.

$A_G = \oplus L^2(G,\pi)$	<u>Hopf algebra</u> (co-algebra)		
$\cap$			
$C(G)$			
$\cap$			
$L^2(G)$	<u>Hilbert algebra</u>	}	convolution
$\cap$			
$L^1(G)$	<u>Banach algebra</u>	}	
$\cap$			
$C^*(G)$	<u>stellar algebra</u>	}	composition of operators in $L^2(G)$
$\cap$			
$\mathcal{U}(G)$	<u>von Neumann algebra</u>	}	

EXERCISES

1. *Let* H_1 *and* H_2 *be two Hilbert spaces. Prove that any operator* A *in* $H_1 \otimes H_2$ *which commutes to all operators* $T \otimes 1$ $(T \in \mathrm{End}\, H_1)$ *can be written in the form* $1 \otimes B$ *for some* $B \in \mathrm{End}\, H_2$.
(Hint: Introduce an orthonormal basis (e_i) *of* H_1 *and write* A *as a matrix of blocks with respect to this basis*

$$A(e_j \otimes x) = \sum_i e_i \otimes A^i_j x \qquad (A^i_j \in \mathrm{End}\, H_2).$$

Using the commutations $(P_j \otimes 1)\, A = A\, (P_j \otimes 1)$ *where* P_j *is the orthogonal projector on* $\mathbb{C}e_j$, *conclude that* $A^i_j = 0$ *for* $i \neq j$. *Finally, using the commutation relations of* A *with the operators* $U_{ji} \otimes 1$

$$U_{ji}(e_i) = e_j \,, \quad U_{ji}(e_k) = 0 \text{ for } k \neq i \,,$$

conclude that $A^i_i = B \in \mathrm{End}\, H_2$ *is independent of* i . *)*

2. *Generalize the result of the preceding exercise in the following context. Let* (V_i) *and* (W_i) *be two sequences of finite dimensional Hilbert spaces and let* H *be the Hilbert sum*

$$H = \widehat{\bigoplus}\, (V_i \otimes W_i) \,.$$

Prove that all operators $B \in \mathrm{End}(H)$ *which commute to all decomposable operators* $\widehat{\bigoplus}(A_i \otimes 1)$ $(A_i \in \mathrm{End}(V_i)$ *and* $\sup \|A_i\| < \infty)$ *are decomposable and have the form*

$$B = \widehat{\bigoplus}(1 \otimes B_i) \quad \text{for some } B_i \in \mathrm{End}(W_i) \text{ with } \sup \|B_i\| < \infty \,.$$

3. *Check that the formula*

$$f = \sum (\chi_\pi \mid f)\, \chi_\pi \qquad (\text{cf. } (7.6)\,)$$

coïncides with the Fourier inversion formula (end of sec.6) for *invariant* *functions. (Observe that* $\langle \chi , \check{f} \rangle = (\bar{\chi} \mid \check{f}) \overset{!}{=} (\chi \mid f)$ *.)*

4. *Prove that* $L^2(G) * L^2(G) \supset A_G$. *(Hint: if* $f, g \in L^2(G,\pi)$, *check that* $f^* * g = c^f_g$ *as coefficient of the right regular representation.)*

5. Let G be a compact group and K a closed subgroup of G. We denote by A^K the subspace of $L^1(G)$ consisting of functions f such that $f(k_1 g k_2) = f(g)$ ($k_i \in K$: subspace of K-biinvariant functions).

a) Show that A^K is a convolution subalgebra of $L^1(G)$.

b) Let (π, V) be a representation of G. Show that the subspace V^K of K-fixed vectors in V is invariant under all operators

$$\pi(f) \qquad\qquad (f \in A^K).$$

Show that if V is irreducible and $V^K \neq \{0\}$, then V^K is irreducible for $\pi(A^K)$ (observe that we can assume π to be unitary; then if $V^K = W_1 \oplus W_2$, put $V_i = \pi\big(L^1(G)\big)\cdot W_i$ and check that these spaces V_i are invariant under G with $V_1 \perp W_2$, $V_2 \perp W_1$).

c) We assume now that A^K is commutative, V irreducible (hence finite dimensional) and $V^K \neq \{0\}$. By part b, $\dim V^K = 1$ and we fix a $v \in V^K$ with $\|v\| = 1$. In particular, we must have

$$\pi(f)\, v = a_f\, v \quad \text{for some } a_f \in \mathbb{C} \qquad\qquad (f \in A^K).$$

Show that

$$a_f = \int f(x)\, \varphi(x)\, dx = (\check{\varphi} \mid f)$$

with the function $\varphi \in A^K$ defined by

$$\varphi(x) = (v \mid \pi(x)\, v) \quad .$$

If χ is the character of π, show that

$$\varphi(x) = \int_K \chi(xk)\, dk$$

(let P denote the orthogonal projector of V on V^K : $P = \int_K \pi(k)\, dk$, and use

$$\varphi(x) = \mathrm{Tr}(P\, \pi(x)\, P) = \mathrm{Tr}(\pi(x)\, P) \qquad).$$

8 INDUCED REPRESENTATIONS AND FROBENIUS - WEIL RECIPROCITY

We have already used the Hilbert-Schmidt norm of operators in finite dimensional Hilbert spaces (sec.6). We need some analogous results in arbitrary Hilbert spaces. Let thus H_1 and H_2 be two Hilbert spaces and $A \in \mathrm{Hom}(H_1,H_2)$ be a continuous linear operator $H_1 \longrightarrow H_2$. Let us take an orthonormal basis (e_i) of H_1. Then

$\sum \|Ae_i\|^2 \in [0,\infty]$ <u>is independent from the choice of orthonormal basis</u> (e_i) <u>in</u> H_1.

Take indeed an orthonormal basis (ε_j) in H_2 and write Parseval's identity for each vector $Ae_i \in H_2$

$$\|Ae_i\|^2 = \sum_j |(\varepsilon_j \mid Ae_i)|^2 \quad .$$

As the operator A is continuous, it has an adjoint A^* : $(\varepsilon_j \mid Ae_i) = (A^*\varepsilon_j \mid e_i)$ and thus

$$\sum_i \|Ae_i\|^2 = \sum_{i,j} |(A^*\varepsilon_j \mid e_i)|^2 = \sum_j \|A^*\varepsilon_j\|^2 \quad .$$

The last series is obviously independent from the choice of (e_i) !

(8.1) <u>Proposition-definition</u>. <u>An operator</u> $A \in \mathrm{Hom}(H_1,H_2)$ <u>between two Hilbert spaces is a Hilbert-Schmidt operator if for one (hence for all) orthonormal basis</u> $(e_i) \subset H_1$, $\sum \|Ae_i\|^2 < \infty$. <u>In this case we put</u>

$$A \in \mathrm{Hom}_2(H_1,H_2) \;,\quad \|A\|_2^2 = \sum \|Ae_i\|^2 \quad .$$

Let us still fix the orthonormal basis $(e_i) \subset H_1$ and denote by H_1^f the dense subspace of H_1 consisiting of (finite) linear combinations of basic vectors e_i. For any linear operator $A : H_1^f \longrightarrow H_2$ satisfying $\sum \|Ae_i\|^2 < \infty$, we have

$$Ax = \sum x_i Ae_i \quad \text{for every} \quad x = \sum x_i e_i \text{ (finite sum),}$$

hence

$$\|Ax\| \leqslant \sum |x_i| \, \|Ae_i\| \leqslant \left(\sum |x_i|^2\right)^{\frac{1}{2}} \left(\sum \|Ae_i\|^2\right)^{\frac{1}{2}} \quad .$$

Thus

$$\|Ax\| \leqslant \|x\|\|A\|_2 \qquad (x \in H_1^f)$$

and A is bounded (continuous) with a continuous extension to H_1 still satisfying

$$\|Ax\| \leqslant \|A\|_2\|x\| \quad , \quad \|A\| \leqslant \|A\|_2 \quad .$$

From this we infer that there are isomorphisms

$$\begin{array}{ccc} \mathrm{Hom}_2(H_1,H_2) & \longrightarrow & \widehat{\bigoplus_I} H_2 = \ell^2_{H_2}(I) \quad (I = \text{index set of } (e_i)) \\ A & \longmapsto & (Ae_i) \quad . \end{array}$$

As a consequence, $\mathrm{Hom}_2(H_1,H_2)$ is a Hilbert space (it is *complete*). Its scalar product is given by

$$(A \mid B)_2 = \sum (Ae_i \mid Be_i)$$

(and this series converges absolutely when A and B $\in \mathrm{Hom}_2(H_1,H_2)$). Choosing an orthonormal basis (ε_j) in H_2, we could still identify this Hilbert space to $\ell^2(I \times J)$ by

$$\begin{array}{ccc} \mathrm{Hom}_2(H_1,H_2) & \longrightarrow & \ell^2(I \times J) \\ A & \longmapsto & ((\varepsilon_j \mid Ae_i))_{I \times J} \quad . \end{array}$$

(8.2) Lemma. Hilbert-Schmidt operators are compact operators.

Proof. Take $A \in \mathrm{Hom}_2(H_1,H_2)$. As $\sum \|Ae_i\|^2 < \infty$, the set I_o of indices $i \in I$ with $Ae_i \neq 0$ is countable and we can decompose H_1 into the orthogonal sum $H_o \oplus H_o'$ with $H_o' \subset \mathrm{Ker}\, A$. Replacing H_1 by H_o, we can assume H_1 separable if we like. For any finite subset $F \subset I$, let us denote by P_F the orthogonal projector of H_1 on the subspace generated by the e_i , $i \in F$. The continuous operators AP_F have finite rank, hence are compact. But

$$\|A - AP_F\|^2 \leqslant \|A - AP_F\|_2^2 = \sum_{I-F} \|Ae_i\|^2$$

is arbitrarily small when F is sufficiently large. This proves that A is a limit (in uniform norm) of a sequence of compact operators, hence is also compact.

(8.3) Lemma. For $A \in \mathrm{Hom}_2(H_1,H_2)$, $T \in \mathrm{End}(H_1)$ and $S \in \mathrm{End}(H_2)$ we have

$$SAT \in \mathrm{Hom}_2(H_1,H_2) \quad \text{and} \quad \|SAT\|_2 \leqslant \|S\|\|T\|\|A\|_2 \quad .$$

Proof. a) In the special case T = 1, the lemma follows from the definitions

$$\|SA\|_2^2 = \sum \|SAe_i\|^2 \leq \|S\|^2 \sum \|Ae_i\|^2 = \|S\|^2 \|A\|_2^2 .$$

b) But we have noticed (just before (8.1)) that $\|B^*\|_2 = \|B\|_2$ so that, quite generally,

$$\|SAT\|_2 = \|T^*(SA)^*\|_2 \underset{(a)}{\leq} \|T^*\| \|SA\|_2 \leq \|T^*\| \|S\| \|A\|_2 .$$

c) The result follows now from the well known equality $\|T^*\| = \|T\|$.

q.e.d.

Let us denote by $\mathrm{Hom}^f(H_1,H_2)$ the vector space of continuous operators $H_1 \longrightarrow H_2$ *with finite rank*. We have already used the isomorphism

$$H_1^\vee \otimes H_2 \xrightarrow{\sim} \mathrm{Hom}^f(H_1,H_2)$$

$$u \otimes v \longmapsto (x \longmapsto (u \mid x)\, v) .$$

The Hilbert-Schmidt scalar product corresponds to the scalar product

$$\Big(\sum_i x_i \otimes y_i \;\Big|\; \sum_j x'_j \otimes y'_j\Big) = \sum_{i,j} (x_i \mid x'_j)_1 (y_i \mid y'_j)_2$$

(all the sums are finite sums since we consider algebraic tensor products). By completion, we obtain

$$H_1^\vee \hat{\otimes} H_2 \xrightarrow{\sim} \mathrm{Hom}_2(H_1,H_2) .$$

By choice of an orthonormal basis (e_i) in H_1 , elements of $H_1^\vee \hat{\otimes} H_2$ can be represented in a unique way as

$$\sum e_i \otimes y_i \quad \text{with} \quad \sum \|y_i\|^2 < \infty .$$

In fact, the element $\sum e_i \otimes y_i \in H_1^\vee \hat{\otimes} H_2$ corresponds to the operator $A \in \mathrm{Hom}_2(H_1,H_2)$ defined by $Ae_i = y_i$. From (8.3) and (8.2) we see that :

(8.4) Lemma. For any Hilbert space H, $\mathrm{End}_2(H) = \mathrm{Hom}_2(H,H)$ is a two-sided ideal of End(H). Moreover

$$\mathrm{id}_H \in \mathrm{End}_2(H) \iff \dim H < \infty .$$

As a first application of this theory, we prove once more the basic finiteness result for the dimension of irreducible representations of a compact group. The following proof is the most elementary one. It does not use the Peter-Weyl theorem, nor the Stone-Weierstrass theorem, and is applicable to unitary representations (in Hilbert spaces).

(8.5) Proof of the finiteness theorem (5.8), (7.9) in the unitary case (Cartier, Godement). Take a unitary irreducible representation π of G in some Hilbert space H. For $A \in \mathrm{End}(H)$, define

$$\tilde{A} = \int \pi(x)^{-1} A\, \pi(x)\, dx = \int \pi(x)^* A\, \pi(x)\, dx$$

as usual by

$$(u \mid \tilde{A} v) = \int (\pi(x)u \mid A\pi(x)v)\, dx \quad .$$

I claim that if A is a Hilbert-Schmidt operator, $\tilde{A}$ is also a Hilbert-Schmidt operator. To prove this assertion, let (e_i) be an orthonormal basis of H and compute sums over finite subsets $F \subset I$

$$\sum \|\tilde{A} e_i\|^2 = \sum (a_i \mid \tilde{A} e_i) = \sum \int (\pi(x)a_i \mid A\pi(x)e_i)\, dx =$$
$$= \int dx \sum (\pi(x)a_i \mid A\pi(x)e_i) = (1 \mid f) \leqslant \|f\|_2$$

(with $a_i = \tilde{A}e_i$ and the function $f = f_F$ defined by

$$f(x) = \sum (\pi(x)a_i \mid A\pi(x)e_i) \qquad).$$

Using the Cauchy-Schwarz inequality several times

$$\left(\sum \|\tilde{A}e_i\|^2\right)^2 \leqslant \|f\|^2 = \int dx \left|\sum (\pi(x)a_i \mid A\pi(x)e_i)\right|^2 \leqslant$$
$$\leqslant \int dx \sum \|\pi(x)a_i\|^2 \cdot \sum \|A\pi(x)e_j\|^2 \leqslant \sum \|a_i\|^2 \int dx\, \|A\|_2^2 \quad .$$

After simplification by $\sum \|\tilde{A}e_i\|^2 = \sum \|a_i\|^2$ we are left with

$$\sum \|\tilde{A}e_i\|^2 \leqslant \|A\|_2^2 \quad .$$

To finish the proof
we observe that we can take for A the orthogonal projector on a line $\mathbb{C}v$ generated by a $v \in H$ with $\|v\| = 1$: then, the positive function

$$x \longmapsto (\pi(x)\, v \mid A\pi(x)\, v)$$

does not vanish at the origin . Thus $\tilde{A} \neq 0$ in this case. But by construction, the operator $\tilde{A}$ commutes with all $\pi(x)$ $(x \in G)$. Since π is assumed to be irreducible, Schur's lemma (its infinite dimensional version given in (8.6) below) shows that $\tilde{A} = \lambda \cdot 1$ is a non-zero scalar operator. As $\tilde{A}$ must also be a Hilbert-Schmidt operator, we see that H must be finite-dimensional.

(8.6) Schur's lemma. Let H be a Hilbert space, $\Phi \subset \mathrm{End}(H)$ a set of operators containing the adjoints of all its elements. If Φ is (topologically) irreducible in H, the only operators commuting to Φ are the scalar operators : $\Phi' = \mathbb{C} \cdot \mathrm{id}_H$ in End(H) .

Proof. (Since the proof of this lemma involves no group theory, we only sketch it. For more details on spectral theory, cf. S. Lang 1975, Appendix 1, Th.4 p.362 .) Let $T \in \mathrm{End}(H)$ be an operator commuting to all operators belonging to Φ . The operators

$$\tfrac{1}{2}(T + T^*) \quad \text{and} \quad \tfrac{1}{2}(T - T^*)/i$$

are hermitian and still commute to Φ (because $A \in \Phi$ implies $A^* \in \Phi$). Denote by A resp. B these operators, so that $T = A + iB$. Since the spectral projectors of A and B are weak limits of polynomials in A (resp. B), these spectral projectors still belong to the commutor Φ' of Φ (this commutor Φ' is the intersection of the kernels of the weakly continuous mappings $S \mapsto ST - TS$, $T \in \Phi$). By irreducibility assumption of Φ , these projectors can only be 0 or 1, and A (resp. B) is a multiple of the identity operator. Observe that $\Phi' = \mathbb{C}\, \mathrm{id}_H$ *characterizes* (topological) irreducibility of Φ in H (orthogonal projectors on invariant subspaces commute to Φ). q.e.d.

INDUCTION

Let G be a compact group and K a closed (hence compact) subgroup of G. We are going to decompose the left regular representation of G in $L^2(G/K)$ (this representation will turn out to be induced from the identity representation of K) as an introduction to the study of more general induced representations. Observe that we studied such a problem in sec.3 with $G = SO_3(\mathbb{R})$ and $K = SO_2(\mathbb{R})$. The results of sec.5 allow us to generalize the situation of sec.3 as follows. The *right* regular representation of G admits a decomposition of the form

$$\rho \cong \widehat{\bigoplus}(1 \otimes \pi) \quad \text{in} \quad L^2(G) = \widehat{\bigoplus}(V_\pi^\vee \otimes V_\pi) \quad .$$

Since the subspace $L^2(G/K) \subset L^2(G)$ consists of the functions which are fixed under right translations coming from elements of K,

$$L^2(G/K) = \widehat{\bigoplus}(V_\pi^\vee \otimes V_\pi^K) \quad .$$

In this isomorphism, left translations of G still act through the first factors of the tensor products

$$\ell_K = \widehat{\bigoplus}\, \check{\pi} \otimes 1_{V^K} \quad \text{in} \quad L^2(G/K) \cong \widehat{\bigoplus}(V_\pi^\vee \otimes V_\pi^K) \quad .$$

Thus the multiplicity of the representation $\check{\pi}$ in ℓ_K is equal to the dimension of V_π^K (subspace of K-invariants in V_π). But the restriction $\pi|_K$ of π in V_π can be written as $\dot{\oplus}\, m_i \pi_i$ with a sum over the unitary dual $\hat{K}$ of K and thus $\check{\pi}|_K = \oplus\, m_i \check{\pi}_i$ with the same multiplicities m_i. In particular for the identity representation in dimension 1 of K, we get

$$m_o = \dim V_\pi^K = \dim V_{\check{\pi}}^K .$$

Thus the multiplicity of $\check{\pi}$ in ℓ_K is m_o and replacing $\check{\pi}$ by π we can state

<u>the multiplicity of</u> π <u>in</u> ℓ_K <u>is equal to the multiplicity of</u> id_K <u>in</u> $\pi|_K$.

Before giving the general definition of induced representations, we have to review some results on measures. The canonical projection $G \xrightarrow{p} G/K$ is a *proper map* and the Haar measure ds of G has an *image* $d\dot{x}$ on G/K characterized by

$$\mu(f) = \int_{G/K} f(\dot{x})\, d\dot{x} = \int_G f(p(s))\, ds \qquad (f \in C(G/K))$$

(8.7) <u>Lemma</u>. <u>Negligible sets of</u> G/K (<u>relative to the measure</u> dx) <u>are those sets</u> N <u>for which</u> $p^{-1}(N)$ <u>is negligible in</u> G (<u>relative to the Haar measure</u> ds <u>of</u> G). <u>Moreover</u>, <u>for any</u> $f \in C(G)$ (<u>or by extension any</u> $f \in L^1(G)$)

$$\int_{G/K} d\dot{x} \int_K f(xk)\, dk = \int_G f(x)\, dx .$$

<u>The measure</u> $d\dot{x}$ <u>is invariant under left translations from</u> G <u>in</u> G/K .

<u>Proof</u>. By definition of $d\dot{x}$, we have

$$\int_{G/K} d\dot{x} \int_K f(xk)\, dk = \int_G \int_K f(xk)\, dk\, dx .$$

Using Fubini's theorem, this integral is also

$$\int_K dk \int_G f(xk)\, dx = \int_K dk \int_G f(x)\, dx = \int_G f(x)\, dx .$$

The other statements of the lemma result from this formula (take for f the characteristic function of a negligible set $N \subset G/K$ or replace $f \in C(G)$ by one of its left translates). q.e.d.

Let now (σ, V) be a unitary representation of K. We define the Hilbert space $L^2(G,V)$ as the completion of the space $C(G,V)$ of continuous functions $G \to V$ with the norm

$$\|f\|^2 = \int_G \|f(x)\|^2 \, dx \quad .$$

The elements

$$f \in L^2(G,V) \text{ such that } f(kx) = \sigma(k)\, f(x) \text{ for all } k \in K$$

constitute a subspace $H \in L^2(G,V)$. Since $\|f(x)\|$ only depends on the coset Kx of x for f in H (σ is assumed to be unitary), we define

$$\|f\|_H^2 = \int_{K\backslash G} \|f(x)\|^2 \, dx \quad ,$$

$$(f \mid g)_H = \int_{K\backslash G} (f(x) \mid g(x)) \, dx$$

(for f and g in H : norm and scalar product under the integral sign being computed in V). The induced representation $\rho = \mathrm{Ind}_K^G(\sigma)$ is by definition the right regular representation of G in $H = H_\sigma \subset L^2(G,V)$:

$$f \in H \implies \rho(s)\, f \in H \quad (\text{where } \rho(s)f\,(x) = f(xs)\,) \quad .$$

This induced representation is unitary. For example, if σ is the identity representation of K (in dimension 1), $V = \mathbb{C}$, $L^2(G,V) = L^2(G)$ and H is simply $L^2(K\backslash G)$.

(8.8) <u>Proposition</u>. 1) <u>The linear map</u> $H^G \to V^K$ <u>given by</u> $f \mapsto f(e)$ <u>is an isomorphism of vector spaces</u>.

2) <u>Let</u> π <u>be a unitary representation of</u> G. <u>Then there is an equivalence</u>

$$\pi \otimes \mathrm{Ind}_K^G(\sigma) \xrightarrow{\sim} \mathrm{Ind}_K^G(\pi|_K \otimes \sigma) \quad : \quad H_\pi \hat{\otimes} H \to \tilde{H}$$

<u>given by</u> $v \otimes f \mapsto \varphi$ <u>with</u> $\varphi(x) = \pi(x)v \otimes f(x)$.

<u>Proof</u>. The elements of H^G (G-fixed elements in H) are the functions $f : G \to V$ which are (equal nearly everywhere to a) constant

$$f(x) = f(ex) = \rho(x)f(e) = f(e) \quad .$$

As

$$\rho(k)f(e) = f(k) = f(e) \quad ,$$

we have $f(e) \in V^K$ and part 1) of the proposition results from this.

To check 2), let us first show that the functions φ (as defined in the proposition) belong to the space of the induced representation $\mathrm{Ind}_K^G(\pi/_K \otimes \sigma)$:

$$\varphi(kx) = \pi(kx)v \otimes f(kx) = \pi(k)\pi(x)v \otimes \sigma(k)f(x) =$$
$$= \pi/_K \otimes \sigma\ (k)\ (\varphi(x)) .$$

Next, we show that $v \otimes f \longmapsto \varphi$ is a G-morphism (intertwining $\pi \otimes \rho$ and the right regular representation $\tilde{\rho} = \mathrm{Ind}_K^G(\pi/_K \otimes \sigma)$) :

$$\pi \otimes \rho\ (s)\ (v \otimes f) = \pi(s)v \otimes \rho(s)f$$

is mapped on the function Φ defined by

$$\Phi(x) = \pi(x)\pi(s)v \otimes (\rho(s)f)(x) = \pi(xs)v \otimes f(xs) =$$
$$= (\tilde{\rho}(s)\varphi)(x)$$

as desired. Now, we check that $v \otimes f \longmapsto \varphi$ is isometric (hence injective). If (e_i) is an orthonormal basis of H_π, every element of $H_\pi \hat{\otimes} H$ can be written uniquely $\sum e_i \otimes f_i$ with $\sum \|f_i\|^2 < \infty$, and such an element has for image the function $\varphi = \sum \varphi_i$: $x \longmapsto \sum \pi(x)e_i \otimes f_i(x)$ having norm

$$\|\varphi\|^2_{\tilde{H}} = \int_{K\backslash G} \|\varphi(x)\|^2\, d\dot{x} = \int_{K\backslash G} \|\sum \pi(x)e_i \otimes f_i(x)\|^2\, d\dot{x} =$$
$$= \int_{K\backslash G} \sum \|f_i(x)\|^2\, d\dot{x} = \sum \int_{K\backslash G} \|f_i(x)\|^2\, d\dot{x} =$$
$$= \sum \|f_i\|^2_H = \|\sum e_i \otimes f_i\|^2$$

(the third equality is justified by the fact that $(\pi(x)e_i)$ is also an orthonormal basis of H_π since π is unitary). Finally, to see that $v \otimes f \longmapsto \varphi$ is onto, it is enough to see that all continuous functions $\Phi \in \tilde{H}$ belong to the image. But write the (unique) expression of $\Phi(x)$ using the orthonormal basis $(\pi(x)e_i)$ of H_π

$$\Phi(x) = \sum \pi(x)e_i \otimes f_i(x) \qquad (f_i(x) \in V)$$
$$\|\Phi(x)\|^2 = \sum \|f_i(x)\|^2 .$$

Therefore

$$\sum \pi(kx)e_i \otimes f_i(kx) = \Phi(kx) = \pi(k) \otimes \sigma(k) \cdot \Phi(x) =$$
$$= \sum \pi(k)\pi(x)\, e_i \otimes \sigma(k)\ f_i(x) .$$

The uniqueness of the decomposition gives $f_i(kx) = \sigma(k)f_i(x)$, i.e. $f_i \in H$ and this concludes the proof.

Let us observe that if we consider $\mathbb{C}$ as trivial G- or K-space the isomorphism of the preceding proposition can be written as

$$\mathrm{Hom}_G(\mathbb{C},H) \xrightarrow{\sim} \mathrm{Hom}_K(\mathbb{C},V) \quad .$$

This form admits the following generalization.

(8.9) Reciprocity theorem (Frobenius-Weil). Let (π,H_π) be a unitary representation of a compact group G and (σ,V) a unitary representation of one of its closed subgroup K. Put $\rho = \mathrm{Ind}_K^G(\sigma)$ and $H = H_\rho$. Then there is a canonical isomorphism

$$\mathrm{Mor}_G(H_\pi,H) \xrightarrow{\sim} \mathrm{Mor}_K(H_\pi,V)$$

where we take for morphisms between two representation spaces, the Hilbert-Schmidt morphisms.

Proof. We have already seen (sec.6) that in the identification

$$H_\pi^\vee \otimes H_\rho \xrightarrow{\sim} \mathrm{Hom}_2(H_\pi,H_\rho)$$

the representation $\check{\pi}\otimes\rho$ is transformed into the representation

$$A \longmapsto \rho(s)\,A\,\pi(s)^{-1} \qquad (s \in G,\ A \in \mathrm{Hom}(H_\pi,H_\rho)).$$

At least, we have seen this for *finite dimensional* representations. But for infinite dimensional unitary representations, we can complete the algebraic tensor product of Hilbert spaces and get an isomorphism with the space of Hilbert-Schmidt operators (observation after (8.3))

$$H_\pi^\vee \hat{\otimes} H_\rho \xrightarrow{\sim} \mathrm{Mor}(H_\pi,H_\rho) \quad .$$

Thus G-morphisms $H_\pi \longrightarrow H_\rho$ correspond to G-invariants in $H_\pi^\vee \hat{\otimes} H_\rho$. Thus we have

$$\mathrm{Mor}_G(H_\pi,H_\rho) \cong (H_\pi^\vee \hat{\otimes} H_\rho)^G$$

and since $H_\pi^\vee \hat{\otimes} H_\rho = \tilde{H}$ can be identified with the space of the induced representation $\pi|_K \otimes \sigma$ (of K in $\tilde{V}$) we infer

$$(H_\pi^\vee \hat{\otimes} H_\rho)^G \cong \tilde{H}^G \cong \tilde{V}^K = (H_\pi^\vee \hat{\otimes} V)^K \quad .$$

The conclusion follows then from the similar identity

$$(H_\pi^\vee \hat{\otimes} V)^K \cong \mathrm{Mor}_K(H_\pi,V) \qquad \text{q.e.d.}$$

(8.10) Corollary. Let $\pi \in \hat{G}$ and $\sigma \in \hat{K}$. Then

multiplicity of π in $\mathrm{Ind}_K^G(\sigma)$ = multiplicity of σ in $\pi|_K$.

Proof. Let us still denote by H the (infinite dimensional in general) space of $\mathrm{Ind}_K^G(\sigma)$. Any G-morphism $H_\pi \to H$ must send H_π into the isotypical component of π in H : this isotypical component is isomorphic to a $\oplus_I H_\pi$ whence

$$\mathrm{Mor}_G(H_\pi, \oplus_I H_\pi) = \oplus_I \mathrm{Mor}_G(H_\pi, H_\pi) = \oplus_I \mathbb{C}$$

(by Schur's lemma). On the other hand, every K-morphism $H_\pi \to V$ must vanish on the isotypical components of H_π which are not equivalent to (or disjoint from) σ. Let us write the isotypical component of σ in H_π as $V \otimes \mathbb{C}^m$:

$$\mathrm{Mor}_K(H_\pi, V) = \mathrm{Mor}_K(V \otimes \mathbb{C}^m, V) \xrightarrow{\sim} \mathrm{Mor}(\mathbb{C}^m, \mathrm{Mor}_K(V,V)) .$$

The last isomorphism associates to $A : V \otimes \mathbb{C}^m \to V$ the morphism

$$v \mapsto \text{restriction of } A \text{ to the subspace } V \otimes \mathbb{C}v \cong V .$$

Since $\mathrm{Mor}_K(V,V) = \mathbb{C}\, 1_V$ (Schur's lemma), we have

$$\mathrm{Mor}_K(H_\pi, V) \cong \text{dual of } \mathbb{C}^m .$$

The isomorphism of the theorem implies equality of the dimension of the spaces. They are respectively

$$\mathrm{Card}(I) = \text{multiplicity of } \pi \text{ in } \mathrm{Ind}_K^G(\sigma) ,$$
$$m = \text{multiplicity of } \sigma \text{ in } \pi|_K .$$ q.e.d.

Observe that the statement of the theorem can be written

$$\mathrm{Mor}_G(\pi, \mathrm{Ind}_K^G(\sigma)) \xrightarrow{\sim} \mathrm{Mor}_K(\pi|_K, \sigma) .$$

In the language of categories, the *functors "induction from K to G" and "restriction from G to K"* are *adjoint* from each other.

GEOMETRIC INTERPRETATION OF INDUCTION

Let us just indicate how induction can be interpreted with vector bundles (for details on this theory, cf. Dieudonné 1970, Chap.XVI). If K is still a closed subgroup of the compact group G, a representation (σ, V) of K defines a vector bundle over $B = K\backslash G$ as follows. The total space of this bundle is the set of K orbits in $V \times G$ with respect to the K-action

$$k \cdot (v,x) = (\sigma(k)\, v , kx) .$$

Here is a diagram of this fibration

$$\begin{array}{l} E \;=\; V \overset{K}{\times} G \;=\; V \times G/\sim \text{ where } (v,x) \sim (\sigma(k)v, kx) \\ \big\downarrow \\ B \;=\; K\backslash G \ni \dot{x} \quad \text{(equivalence class for } x \sim kx) \quad . \end{array}$$

Sections of this bundle are identified with functions

$$f : x \longmapsto f(x) = v \quad \text{with} \quad f(kx) = \sigma(k)v = \sigma(k)\, f(x) \quad .$$

These are precisely the elements of the space of the induced representation. In other words, the space H of the induced representation $\mathrm{Ind}_K^G(\sigma)$ *is* the Hilbert space of square summable sections of the bundle $E \longrightarrow K\backslash G$ associated to (σ, V) and the induced representation acts by right translations. On the basis $K\backslash G$, the action is $x \longmapsto k(x) = xk^{-1}$ and on sections (identified to functions $G \longrightarrow V$) the equivariant action is

$$_s f(x) = f(s^{-1}(x)) = f(xs) = \rho(s) f(x) \quad .$$

Of course, one could define the induced representation (ρ, H) of (σ, V) by using functions $f \in L^2(G,V)$ satisfying

$$f(xk) = \sigma(k)^{-1}(f(x)) \quad .$$

With this other definition of induced representation, we would consider the vector bundle with total space

$$G \overset{K}{\times} V = G \times V/\sim \quad \text{where} \quad (x, v) \sim (xk^{-1}, \sigma(k)\, v) \quad .$$

Denoting by $x \cdot v = [x,v]$ the equivalence class of (x,v), we see that this "product" has the typical property $xk \cdot v = x \cdot \sigma(k)\, v$ (comparable to tensor products over K !) .

EXERCISES

1. Let H be a Hilbert space and I any index set. Prove that $\ell^2(I) \hat{\otimes} H$ and $\ell^2_H(I)$ are canonically isomorphic. More generally, if X is any measure space, prove that $L^2(X) \hat{\otimes} H$ and $L^2(X,H)$ are canonically isomorphic.

2. Give a proof of the second part of (8.8) along the following lines. Using Ex.1 above and sec.5, write

$$L^2(G,V) = L^2(G) \hat{\otimes} V = \widehat{\bigoplus_{\tau\in\hat{G}}} V_\tau^\vee \otimes V_\tau \otimes V .$$

Consider this space $L^2(G,V)$ as a representation space of G with

$$\lambda(s)f(x) = \sigma(s)\, f(s^{-1}x) ,$$
$$\rho(t)f(x) = f(xt) .$$

Then the space of $\mathrm{Ind}_K^G(\sigma)$ consists of the $\lambda(K)$-invariants in $L^2(G,V)$. Write this space of invariants for $\mathrm{Ind}_K^G(\pi|_K \otimes \sigma)$ as

$$\bigoplus_\tau \mathrm{Hom}_K(V_\tau , V_\pi \otimes V) \otimes V_\tau$$

and conclude by showing that in this expression, the space V_π can be "pulled in front of the sum".

3. Let (π,H) and (π',H') be two unitary representations of a group G. Assume that there is a (continuous) bijective operator $A : H \longrightarrow H'$ such that

$$A\,\pi(s) = \pi'(s)\,A \quad \text{for all } s \in G .$$

Prove that there is a unitary operator $B : H \longrightarrow H'$ (i.e. B is a bijective isometry) giving a unitary equivalence of π and π'. (Hint: Define $T = (A^*A)^{-\frac{1}{2}}$ and put $B = A\,T$. This exercise is taken from Borel 1972, (5.2) p.46 .)

9 TANNAKA DUALITY

Let G be a compact group. We consider the *category* $\mathcal{C}_G$ of (complex) finite dimensional representations of G : its *objects* are the finite dimensional representations (π, V) of G and its *morphisms* between two such representations are the G-morphisms

$$\mathrm{Mor}(\pi,\sigma) = \mathrm{Hom}_G(V_\pi, V_\sigma) \quad .$$

Now consider a *fixed* element $s \in G$ and the corresponding collection of operators $\pi(s)$ when π or (π, V) runs over all objects of $\mathcal{C}_G$.
By definition of G-morphisms, for π and $\sigma \in \mathcal{C}_G$, $A \in \mathrm{Mor}(\pi,\sigma)$, the following diagram is *commutative*

$$\begin{array}{ccc} V & \xrightarrow{\pi(s)} & V \\ \downarrow A & & \downarrow A \\ W & \xrightarrow{\sigma(s)} & W \end{array} \qquad (V = V_\pi \ , \ W = V_\sigma).$$

Still by definition

$$\pi \otimes \sigma\ (s) = \pi(s) \otimes \sigma(s) \quad ,$$

$$\overline{\pi(s)} = \bar{\pi}(s) \quad \text{for } \pi \text{ unitary in } \mathcal{C}_G \ .$$

We call <u>representation</u> of the category $\mathcal{C}_G$ any family (or collection !) of endomorphisms

$$(\gamma_V) \quad \text{or} \quad (\gamma_\pi) \qquad\qquad (\gamma \in \mathrm{End}(V_\pi))$$

(parametrized by the class of objects of $\mathcal{C}_G$) having the above three properties. <u>Axiomatically</u>, representations of $\mathcal{C}_G$ are collections (γ_V) satisfying

1. <u>For</u> $A \in \mathrm{Mor}(\pi,\sigma)$, $V = V_\pi$ <u>and</u> $W = V_\sigma$ <u>the following diagram is commutative</u>

$$\begin{array}{ccc} V & \xrightarrow{\gamma_V} & V \\ \downarrow A & & \downarrow A \\ W & \xrightarrow{\gamma_W} & W \end{array} \quad .$$

2. $\gamma = (\gamma_V)$ is multiplicative in the sense that

$$\gamma_{\pi \otimes \sigma} = \gamma_\pi \otimes \gamma_\sigma \qquad (\pi, \sigma \in \mathcal{C}_G).$$

3. For π unitary in $\mathcal{C}_G$, $\gamma_{\bar{\pi}} = \overline{\gamma_\pi}$.

Let us draw a few consequences from these axiomatic properties of representations of $\mathcal{C}_G$.

a) Let $(\pi_0, \mathbb{C})$ denote the identity representation of G in dimension 1. Then $\gamma_0 = \gamma_{\pi_0} = \mathrm{id}_{\mathbb{C}}$.

Indeed, γ_0 must be given by scalar multiplication by a certain $\lambda \in \mathbb{C}$ and property 2 above gives

$$\gamma_0 = \gamma_{\pi_0 \otimes \pi_0} = \gamma_0 \otimes \gamma_0 ,$$

hence $\lambda^2 = \lambda$ and $\lambda = 1$.

b) If (π, V) and (σ, W) are two elements of $\mathcal{C}_G$, then

$$\gamma_{V \oplus W} = \gamma_V \oplus \gamma_W .$$

To prove this, consider the G-morphisms $V \longrightarrow V \oplus W$ and $W \longrightarrow V \oplus W$ and the corresponding commutative diagrams

$$\begin{array}{ccc} V & \xrightarrow{\gamma_V} & V \\ \downarrow & & \downarrow \\ V \oplus W & \xrightarrow[\gamma_{V\oplus W}]{} & V \oplus W \end{array} , \qquad \begin{array}{ccc} W & \xrightarrow{\gamma_W} & W \\ \downarrow & & \downarrow \\ V \oplus W & \xrightarrow[\gamma_{V\oplus W}]{} & V \oplus W \end{array} .$$

Commutativity implies that $\gamma_{V\oplus W}$ leaves $V \cong V \oplus \{0\}$ (and similarly W) invariant and induces γ_V (resp. γ_W) in it. Thus $\gamma_{V\oplus W} = \gamma_V \oplus \gamma_W$.

c) For $\pi \in \mathcal{C}_G$, $\gamma_{\check{\pi}} = \check{\gamma}_\pi$.

Let us identify $V_{\check{\pi}} \otimes V_\pi$ to $\mathrm{End}(V_\pi)$ in the usual way. Then the representation $\check{\pi} \otimes \pi$ is transformed into the representation $A \longmapsto \pi(x) \cdot A \cdot \pi(x)^{-1}$. This representation leaves the scalar operators fixed, whence a G-morphism from the identity representation $(\pi_0, \mathbb{C})$ in dimension 1 into $(\check{\pi} \otimes \pi, \mathrm{End}\, V_\pi)$ furnishing commutative diagrams as shown on next page. Using property 2 for $\gamma_{\check{\pi} \otimes \pi} = \gamma_{\check{\pi}} \otimes \gamma_\pi$ acting by

$$A \longmapsto \gamma_\pi \circ A \circ {}^t\gamma_{\check{\pi}} \qquad \text{(Ex.1.c of sec.5)}$$

Commutativity of the following diagram gives $\mathrm{id} = \gamma_\pi \cdot {}^t\gamma_{\check{\pi}}$
($A = \mathrm{id.}$ of V_π corresponds to $1 \in \mathbb{C}$)

$$\gamma_{\check{\pi}} = {}^t\gamma_\pi^{-1} = \check{\gamma}_\pi .$$

$$\begin{array}{ccc}
\mathbb{C} & \xrightarrow[\gamma_0 = id_{\mathbb{C}}]{} & \mathbb{C} \\
\downarrow & & \downarrow \\
\mathrm{End}\, V_\pi & \longrightarrow & \mathrm{End}\, V_\pi \\
\wr\uparrow & & \wr\uparrow \\
V_\pi^\vee \otimes V_\pi & \xrightarrow{\gamma_{\check{\pi}} \otimes \gamma_\pi} & V_\pi^\vee \otimes V_\pi
\end{array}$$

(Alternatively, if (e_i) is a basis of $V = V_\pi$ and (ε_i) the dual basis of $V^\vee$, the G-morphism $\pi_0 \to \check{\pi} \otimes \pi$ sends $1 \in \mathbb{C}$ on $\sum \varepsilon_i \otimes e_i$ and the commutativity of the diagram requires that

$$\sum \varepsilon_i \otimes e_i = \gamma_{\check{\pi}} \otimes \gamma_\pi \sum \varepsilon_i \otimes e_i = \sum \gamma_{\check{\pi}}(\varepsilon_i) \otimes \gamma_\pi(e_i).$$

But the dual basis $(\eta_i) = (\check{\gamma}_\pi(\varepsilon_i))$ of $(\gamma_\pi(e_i))$ is characterized uniquely by

$$\sum \varepsilon_i \otimes e_i = \sum \eta_i \otimes \gamma_\pi(e_i) ,$$

hence $\gamma_{\check{\pi}}(\varepsilon_i) = \check{\gamma}_\pi(\varepsilon_i)$ and thus $\gamma_{\check{\pi}} = \check{\gamma}_\pi$.)

d) <u>If π is unitary, γ_π is also unitary.</u>

By axiom 3, $\gamma_{\bar{\pi}} = \bar{\gamma}_\pi \in \mathrm{End}(\bar{V}_\pi)$. But in the usual identification $\bar{V} \cong V^\vee$ (given by Riesz' theorem), we have $\bar{\pi} = \pi^\vee$ (Ex.2 of sec.2). Hence

$$\bar{\gamma}_\pi = \gamma_{\check{\pi}} = \check{\gamma}_\pi \quad , \quad \gamma_\pi^{-1} = {}^t\bar{\gamma}_\pi = \gamma_\pi^* \quad .$$

Now the set of representations $\mathrm{Rep}(\mathcal{C}_G)$ of $\mathcal{C}_G$ is a group with respect to composition of the endomorphisms γ_V and we have a homomorphism

$$G \to \mathrm{Rep}(\mathcal{C}_G) : s \mapsto (\pi(s))_{\pi \in \mathcal{C}_G} .$$

In fact, with the topology of simple (pointwise) convergence, $\mathrm{Rep}(\mathcal{C}_G)$ is a topological group. Since any finite dimensional representation of G is unitarizable and a sum of irreducible representations, there is a continuous homomorphism

$$\mathrm{Rep}(\mathcal{C}_G) \to \prod_{\hat{G}} U(V_\pi) : (\gamma_\pi)_{\mathcal{C}_G} \mapsto (\gamma_\pi)_{\pi \in \hat{G}}$$

(points b and d above). The image is a closed, hence compact subgroup of the product with which we identify $\mathrm{Rep}(\mathcal{C}_G)$. The continuous homomorphism

$$G \longrightarrow \mathrm{Rep}(\mathcal{C}_G) \quad : \quad s \longmapsto \gamma^s = (\gamma_V^s) = (\pi(s))$$

is injective by the Peter-Weyl theorem (first theorem of sec.4) hence a homeomorphism onto its image. In fact, it is surjective :

Theorem (Tannaka). The canonical homomorphism

$$G \longrightarrow \mathrm{Rep}(\mathcal{C}_G) \quad : \quad s \longmapsto \gamma^s$$

where $\gamma_V^s = \pi(s)$ $(V = V_\pi)$ is an isomorphism of topological groups.

To prove this theorem, we need a lemma.

Lemma. Let V be the space of a finite dimensional representation π of G. If $v \in V^G$ is a fixed vector (i.e. $\pi(s)v = v$ for all $s \in G$), then

$$\gamma_V(v) = v \quad \text{for all} \quad \gamma \in \mathrm{Rep}(\mathcal{C}_G) .$$

Proof of the lemma. By hypothesis $v \in V^G$, we can define a G-morphism from the identity $(\pi_o, \mathbb{C})$ in dimension 1 to (π, V) by sending 1 on v. Thus we get a commutative diagram

$$\begin{array}{ccc} 1 \in \mathbb{C} & \xrightarrow{\ id\ } & \mathbb{C} \\ \downarrow \quad \downarrow & & \downarrow \\ v \in V & \xrightarrow{\ \gamma_V\ } & V \end{array}$$

whence the conclusion !

Proof of the theorem. The elements of the image of $G \longrightarrow \mathrm{Rep}(\mathcal{C}_G)$ are the γ such that there exists an $s \in G$ with $\gamma_\pi = \pi(s)$ (all $\pi \in \mathcal{C}_G$). By negation

$$\gamma \notin \mathrm{Image}(G) \iff \begin{cases} \text{for every } s \in G \text{ there exists } (\pi, V) \in \mathcal{C}_G \\ \text{such that } \quad \gamma_V \neq \pi(s) \quad . \end{cases}$$

We show that the existence of such a $\gamma \notin \mathrm{Image}(G)$ leads to a contradiction. But a condition $\gamma_V \neq \pi(x)$ defines an open set in G containing the element s. Since G is compact, we can find a finite set of elements s_i such that the corresponding open sets cover G. But these elements s_i correspond to a finite set of representations (π_i, V_i) of G. We consider $W = \bigoplus V_i$ (finite sum of finite dimensional representations) and $\gamma_W = \bigoplus \gamma_{W_i}$ (point b above). The property $\gamma \notin \mathrm{Image}(G)$ would imply the existence of $\sigma = \bigoplus \pi_i$ with $\gamma_W \notin \sigma(G) \subset \mathrm{Gl}(W)$. The two compact sets

$\sigma(G)$ and $\sigma(G)\gamma_W$ would be disjoint and it would be possible to construct a continuous function taking value 0 on the first and value 1 on the second set. Approximating such a function uniformly by a polynomial function P on End(W), we would find a polynomial with

$$|P| \leqslant 1/3 \quad \text{on} \quad \sigma(G) \;, \qquad P \geqslant 2/3 \quad \text{on} \quad \sigma(G)\gamma_W \,.$$

Averaging P on G using the Haar measure, we would thus construct the polynomial Q (of degree smaller or equal to the degree of P)

$$Q(A) = \int_G P(\sigma(t)\, A)\, dt \quad .$$

This polynomial Q would have the properties

$$Q(A) = Q(\sigma(s)\, A) \quad (\text{all } s \in G), \quad Q(A) \neq Q(\gamma_W A) \,.$$

But the finite dimensional representation of G in the space of polynomials on End(W) of degree smaller or equal to deg(P) would contradict the assertion of the lemma (with the fixed vector v = Q !). q.e.d.

10 REPRESENTATIONS OF THE ROTATION GROUP

In this section, we consider the compact group $G = SO_3(\mathbb{R})$ of proper rotations in $\mathbb{R}^3$. Its elements are the 3×3 real orthogonal matrices with determinant 1. We first have to give its Haar measure and study various classical isomorphisms.

We start by reviewing the parametrization of this group by <u>Euler angles</u> (like in any book on mechanics...). Any rotation can be decomposed into : a rotation around the z-axis, a swing around the line of nodes (image of the x-axis under the first rotation) and finally a rotation around around the new z'-axis (cf. picture below). Let us denote by

K = subgroup of rotations around the z-axis,

A = subgroup of rotations around the x-axis

(both subgroups are isomorphic to the circle group $SO_2(\mathbb{R})$: observe that the subgroup now denoted by A was called K in sec.3). Thus we write

$$g = g_{\varphi,\theta,\psi} = \tilde{g}_\psi \, \tilde{h}_\theta \, g_\varphi$$

with

$$g_\varphi \in K \ , \quad \tilde{h}_\theta = g_\varphi h_\theta g_\varphi^{-1} \in g_\varphi A \, g_\varphi^{-1} \ ,$$
$$\tilde{g}_\psi = \tilde{h}_\theta \, g_\psi \tilde{h}_\theta^{-1} \in \tilde{h}_\theta \, K \, \tilde{h}_\theta^{-1} \quad .$$

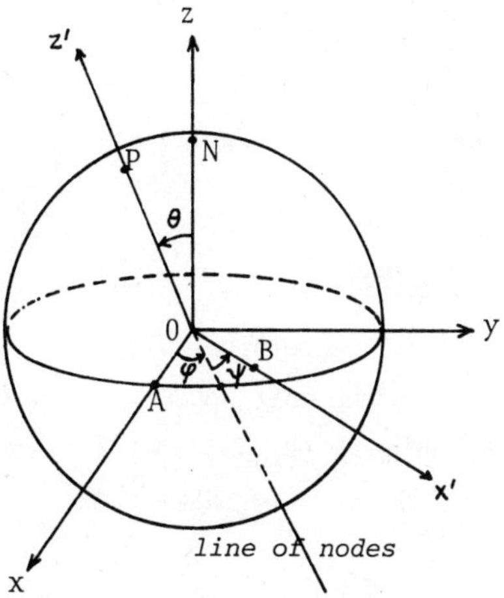

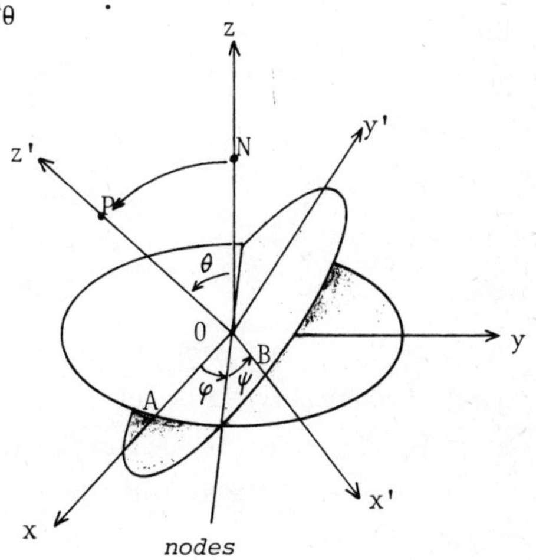

Multiplying throughout, an elementary computation shows that

$$g = g_{\varphi,\theta,\psi} = g_\varphi h_\theta g_\psi \in KAK \ .$$

(Observe that the order of the angles φ and ψ is thus reversed, and we have now rotations around *fixed axes* Ox in A, resp. Oz in K !)
In particular, we have a decomposition $G = KAK$.

Since the Haar measure dx of G satisfies

$$\int_G f(x)\,dx = \int_{G/K} d\dot{x} \int_K f(xk)\,dk$$

with the invariant measure $d\dot{x}$ on G/K identified to the normalized invariant measure $d\Omega$ on this sphere, we have

$$\begin{cases} dg_{\varphi,\theta,\psi} = \frac{1}{4\pi}\, d\Omega\, \frac{1}{2\pi}\, d\psi = \frac{1}{8\pi^2} \sin\theta\, d\varphi\, d\theta\, d\psi \\ (\varphi, \theta, \psi) \in (0,2\pi) \times (0,\pi) \times (0,2\pi) \ . \end{cases}$$

UNIVERSAL COVERING OF SO_3 AND CLASSICAL ISOMORPHISMS

Before we start representation theory, we study the universal covering of the rotation group $G = SO_3(\mathbb{R})$. This group G can be identified to the group $SU_2(\mathbb{C})$ consisting of 2×2 complex matrices g with

$$g^* g = 1 \ , \quad \det(g) = 1 \ .$$

As is easily seen, this group consists of matrices

$$g = \begin{pmatrix} u & v \\ -\bar{v} & \bar{u} \end{pmatrix} \ , \quad |u|^2 + |v|^2 = 1 \qquad (u\ , v \in \mathbb{C}) \ .$$

The two complex parameters u and v of g are called the Cayley-Klein parameters.

We shall establish the following isomorphisms

$$\begin{array}{ccccccc} (\tilde{G} =) & S^3 = \mathbb{H}^1 & \xrightarrow[(a)]{\sim} & SU_2(\mathbb{C}) & & \text{(Cayley-Klein)} \\ \downarrow 2{:}1 & \searrow & (b) \searrow & \swarrow (c) & & \\ (G =) & \mathbb{P}^3(\mathbb{R}) & \cong & SO_3(\mathbb{R}) & & \text{(Euler)} \ . \end{array}$$

Here, we have denoted by $\mathbb{H}^1$ the group of (real) quaternions with norm 1 : obviously, this set is identified with the unit sphere in $\mathbb{R}^4 = \mathbb{H}$. The other identification (homeomorphism) $\mathbb{P}^3(\mathbb{R}) \cong SO_3(\mathbb{R})$ can be seen as follows. Let g be a rotation, $\vec{\omega}$ its axis ($\|\vec{\omega}\| = 1$)

and ϕ its angle (counted with the *corkscrew law*!) so that $-\pi < \phi \leqslant \pi$. In this way, the rotation is identified with a point of the ball of radius π (antipodal points being identified). But this (full) ball can be deformed on the northern hemisphere of a sphere S^3 in $\mathbb{R}^4$ as suggested in the picture.

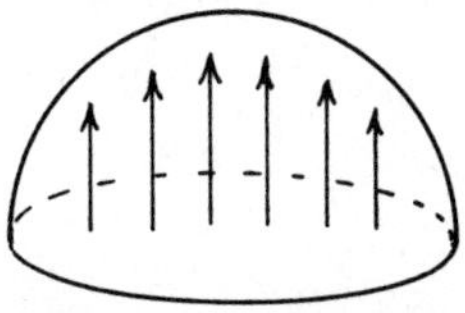

But $\mathbb{P}^3(\mathbb{R})$ is precisely obtained by identification of antipodal points of S^3 so that the homeomorphism between $SO_3(\mathbb{R})$ and $\mathbb{P}^3(\mathbb{R})$ is now clear.

(a) Let us consider the vector space $\mathbb{H}$ of real quaternions as a *complex vector space* with basis 1,j (1,i,j,k being the usual basis of $\mathbb{H}$ over $\mathbb{R}$: $i^2 = j^2 = k^2 = -1$). For this, we consider the scalar multiplications

$$\lambda \circ q = \bar{\lambda} q \qquad (\lambda \in \mathbb{C},\ q \in \mathbb{H} \text{ and product } \bar{\lambda} q \text{ in } \mathbb{H}).$$

Then, for $g \in \mathbb{H}^1$,

$$\sigma_g : q \longmapsto q g^{-1} = q\bar{g}$$

is $\mathbb{C}$-linear and defines a representation (with complex dimension 2) of $\tilde{G} = \mathbb{H}^1$. Obviously, taking successively $g = i, j, k$ $(\in \mathbb{H}^1)$

$$\sigma_i = \begin{pmatrix} i & 0 \\ 0 & -i \end{pmatrix}, \quad \sigma_j = \begin{pmatrix} 0 & 1 \\ -1 & 0 \end{pmatrix}, \quad \sigma_k = \begin{pmatrix} 0 & i \\ i & 0 \end{pmatrix}$$

and finally for $g = u + vj$ $(\det \sigma_g = N(g) = g\bar{g} = 1)$

$$\sigma_g = \begin{pmatrix} u & v \\ -\bar{v} & \bar{u} \end{pmatrix}.$$

Thus we obtain the first isomorphism $\tilde{G} \xrightarrow{\sigma} SU_2(\mathbb{C})$.

(b) The group $\tilde{G} = \mathbb{H}^1$ acts by conjugation on the space E of pure quaternions (E is a *real* subspace of dimension 3 of $\mathbb{H}$)

$$v \in E = \mathbb{R}i + \mathbb{R}j + \mathbb{R}k \subset \mathbb{H}, \quad v = xi + yj + zk,$$

$$\rho(g)v = gvg^{-1} = v', \quad N(v') = N(v) = x^2 + y^2 + z^2 .$$

We obtain in this way a homomorphism $\tilde{G} \longrightarrow O_3(\mathbb{R})$, and since $\tilde{G}$ is connected, its image must be contained in $SO_3(\mathbb{R})$. An easy verification shows that its kernel is the two elements subgroup $\{\pm 1\}$.

(c) Let V be the real vector space (of dimension 3) of 2×2 complex matrices which are hermitian and have zero trace. These matrices can be written

$$X = \begin{pmatrix} x & y+iz \\ y-iz & -x \end{pmatrix} \qquad (\text{hence } -\det X = x^2 + y^2 + z^2).$$

The group $SU_2(\mathbb{C})$ acts in this space by conjugation

$$\tau_g(X) = g X g^* = g X g^{-1}$$

whence a homomorphism

$$\tau : SU_2(\mathbb{C}) \rightarrow SO_3(\mathbb{R})$$

with kernel $\{\pm 1\}$.

The lifting $SO_3(\mathbb{R}) \rightarrow SU_2(\mathbb{C})/\{\pm 1\} \subset Sl_2(\mathbb{C})/\{\pm 1\}$ can also be given in terms of the Euler angles as follows. To determine the image of a rotation

$$g = g_{\varphi,\theta,\psi} = g_\varphi h_\theta g_\psi \mapsto \begin{bmatrix} u & v \\ -\bar{v} & \bar{u} \end{bmatrix} = \pm \begin{pmatrix} u & v \\ -\bar{v} & \bar{u} \end{pmatrix}$$

we consider separately the two cases h_θ and g_φ and use stereographic projection (cf. picture at the bottom of this page).

1) The rotation g_φ is projected on the rotation $\zeta \mapsto e^{i\varphi}\zeta$ of the same angle in the ζ-plane. This rotation is given by the matrix

$$\begin{bmatrix} e^{i\frac{1}{2}\varphi} & 0 \\ 0 & e^{-i\frac{1}{2}\varphi} \end{bmatrix} \in \mathrm{Aut}_{hol}(\mathbb{C}) = Sl_2(\mathbb{C})/\{\pm 1\} .$$

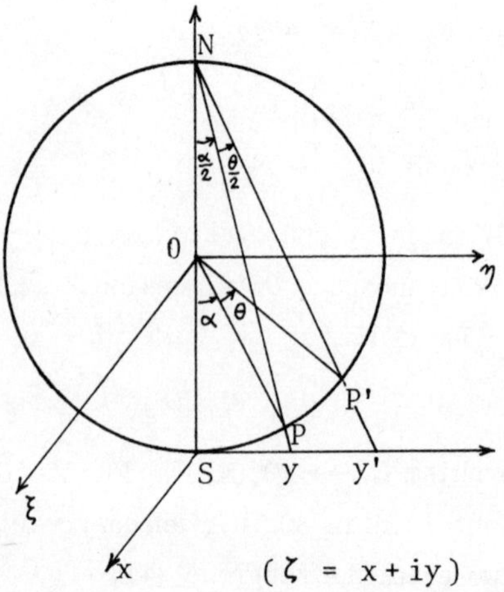

Stereographic projection from the north pole N of a sphere of radius ½, on the tangent plane to the southern pole S .

$(\zeta = x + iy)$

2) The rotation h_θ of angle θ around the axis 0ξ is projected on a homography (or fractional linear transformation) that is completely determined by its action on the imaginary axis Sy in the ζ-plane. If $y = \mathrm{tg}\frac{1}{2}\alpha$, then

$$y' = \mathrm{tg}\,\tfrac{1}{2}(\alpha+\theta) = \frac{\mathrm{tg}\,\frac{1}{2}\alpha + \mathrm{tg}\,\frac{1}{2}\theta}{1 - \mathrm{tg}\,\frac{1}{2}\alpha\;\mathrm{tg}\frac{1}{2}\theta}$$

so that

$$y' = \frac{y + \mathrm{tg}\frac{1}{2}\theta}{-y\,\mathrm{tg}\,\frac{1}{2}\theta + 1} = \frac{\cos\frac{1}{2}\theta\; y + \sin\frac{1}{2}\theta}{i^2\sin\frac{1}{2}\theta\; y + \cos\frac{1}{2}\theta},$$

$$iy' = \frac{\cos\frac{1}{2}\theta\cdot iy + i\sin\frac{1}{2}\theta}{i\sin\frac{1}{2}\theta\cdot iy + \cos\frac{1}{2}\theta}.$$

This proves that

$$\zeta' = \frac{\cos\frac{1}{2}\theta\cdot\zeta + i\sin\frac{1}{2}\theta}{i\sin\frac{1}{2}\theta\cdot\zeta + \cos\frac{1}{2}\theta}$$

And the homography corresponding to h_θ is given by

$$\begin{bmatrix}\cos\frac{1}{2}\theta & i\sin\frac{1}{2}\theta\\ i\sin\frac{1}{2}\theta & \cos\frac{1}{2}\theta\end{bmatrix}.$$

Combining the two results for g_φ and h_θ, we find by matrix multiplication

$$g = g_{\varphi,\theta,\psi} = g_\varphi h_\theta g_\psi \longmapsto \begin{bmatrix}\cos\frac{1}{2}\theta\; e^{i\frac{1}{2}(\varphi+\psi)} & i\sin\frac{1}{2}\theta\, e^{i\frac{1}{2}(\varphi-\psi)}\\ \cdots & \cdots\end{bmatrix}$$

(10.1) Proposition. Let g be a rotation of angle ϕ around a certain axis. Then

$$\cos\tfrac{1}{2}\phi = \mathrm{Re}(u) \text{ in terms of the Cayley-Klein parameters},$$

$$\cos\tfrac{1}{2}\phi = \cos\tfrac{1}{2}\theta\cos\tfrac{1}{2}(\varphi+\psi) \text{ in terms of the Euler angles}.$$

Proof. The angle of rotation is invariant under conjugation (in SO_3 and in SU_2). Consequently, we can find the angle ϕ by diagonalization. Let us find the eigenvalues of the 2×2 matrix with Cayley-Klein parameters u and v :

$$\begin{vmatrix}u-\lambda & v\\ -\bar v & \bar u-\lambda\end{vmatrix} = \lambda^2 - 2\,\mathrm{Re}(u)\lambda + \underbrace{(u\bar u + v\bar v)}_{1} = 0,$$

whence $e^{\pm i\frac{1}{2}\phi} = \lambda_{1,2} = \mathrm{Re}(u) \pm i\sqrt{1-\mathrm{Re}(u)^2}$. q.e.d.

We can use the preceding proposition to determine conjugacy classes in $SO_3(\mathbb{R})$. First in SU_2

$$g = \begin{pmatrix} u & v \\ -\bar{v} & \bar{u} \end{pmatrix} \text{ is conjugate to } \begin{pmatrix} e^{i\frac{1}{2}\phi} & 0 \\ 0 & e^{-i\frac{1}{2}\phi} \end{pmatrix} \text{and to } \begin{pmatrix} e^{-i\frac{1}{2}\phi} & 0 \\ 0 & e^{i\frac{1}{2}\phi} \end{pmatrix}.$$

Thus we can always choose the eigenvalue $\lambda = \lambda_1 = e^{i\frac{1}{2}\phi}$ so that $\mathrm{Im}(\lambda_1) \geqslant 0$ and we fix $\lambda = e^{i\frac{1}{2}\phi}$ with $0 \leqslant \frac{1}{2}\phi \leqslant \pi$, i.e. $0 \leqslant \phi \leqslant 2\pi$. In this way, ϕ is the unique solution of $\cos\frac{1}{2}\phi = \mathrm{Re}(u)$ with $\phi \in [0,2\pi]$. In $SO_3(\mathbb{R}) = SU_2(\mathbb{C})/\{\pm 1\}$ we still have

$$\begin{bmatrix} u & v \\ \cdot & \cdot \end{bmatrix} = \begin{bmatrix} -u & -v \\ \cdot & \cdot \end{bmatrix}$$

so that ϕ and $2\pi - \phi$ lead to conjugate matrices, and we can choose the solution ϕ with $\frac{1}{2}\phi \in [0,\pi/2]$, namely $\phi \in [0,\pi]$. We do this by putting

$$\cos\tfrac{1}{2}\phi = |\mathrm{Re}(u)| = |u_1| \qquad (u = u_1 + iu_2)$$

and by choosing the representation

$$\begin{bmatrix} u & v \\ -\bar{v} & \bar{u} \end{bmatrix} \quad \text{with} \quad 0 \leqslant u_1 \leqslant 1 \quad .$$

Remarks

1) To go from Cayley-Klein parameters to Euler angles, one has to solve

$$|u| = \cos\tfrac{1}{2}\theta \quad (\text{unique } \theta \in [0,\pi[\,),\ \sin\tfrac{1}{2}\theta = |v|$$

and one must have

$$\tfrac{1}{2}(\varphi + \psi) = \arg u \quad , \quad \tfrac{1}{2}(\varphi - \psi + \pi) = \arg v \quad ,$$

whence

$$\varphi = \arg u + \arg v - \tfrac{1}{2}\pi \quad (\mathrm{mod}\ 2\pi)\ ,$$
$$\psi = \arg u - \arg v + \tfrac{1}{2}\pi \quad (\mathrm{mod}\ 2\pi)\ .$$

The closed curve $\varphi \longmapsto g_\varphi$ (parametrized by $\varphi \in [0,2\pi]$) is lifted in $SU_2(\mathbb{C})$ in the (non-closed) curve

$$\varphi \longmapsto \begin{pmatrix} e^{i\frac{1}{2}\varphi} & 0 \\ 0 & e^{-i\frac{1}{2}\varphi} \end{pmatrix}$$

with extremities 1_2 and -1_2 in $SU_2(\mathbb{C})$. This proves that the closed path $\varphi \longmapsto g_\varphi$ is not homotopic to a point in $SO_3(\mathbb{R})$.

2) If $\mathbb{H}$ is regarded as $\mathbb{C}$-vector space with scalar multiplication given by $\lambda \cdot q = q\lambda$, the left regular representation $\mathbb{H}^1 \longrightarrow Gl_2(\mathbb{C})$ is given by

$$\sigma_g = \begin{pmatrix} u & -v \\ \bar{v} & \bar{u} \end{pmatrix} .$$

With the scalar multiplication $\lambda \cdot q = \lambda q$, the right regular representation leads to

$$\sigma_g = \begin{pmatrix} \bar{u} & \bar{v} \\ -v & u \end{pmatrix} .$$

As we see, to insure $\mathbb{C}$-linearity, we have to take group action and scalar multiplication on opposite sides.

3) The subgroup

$$A = \{ g_\varphi : \varphi \in [0, 2\pi] \} \subset SO_3(\mathbb{R})$$

is a *maximal abelian* subgroup. Indeed, any commuting family of rotations (each rotation is a *semi-simple* operator) can be simultaneously diagonalized, hence has a common rotation axis. In particular, all commutative subgroups of $SO_3(\mathbb{R})$ are conjugate to a subgroup of A, and all maximal abelian subgroups of $SO_3(\mathbb{R})$ are conjugate to A.

REPRESENTATIONS

We define a sequence of representations of the universal covering $\tilde{G} = SU_2(\mathbb{C})$ of $G = SO_3(\mathbb{R})$ as follows. For an integer $n \geqslant 0$ (i.e. $n \in \mathbb{N}$), let V^n denote the space of homogeneous polynomials of degree n in two variables (and complex coefficients) :

$$V^o = \mathbb{C} , \quad V^1 = V , \quad \ldots \quad , V^n = \{\text{symmetric tensors in } V^{\otimes n}\}.$$

We define the action of

$$M = \begin{pmatrix} u & v \\ -\bar{v} & \bar{u} \end{pmatrix} \in SU_2(\mathbb{C})$$

on polynomials by *right multiplication*

$$P = P(z_1, z_2) \mapsto P_M = P((z_1, z_2) \cdot M) = $$
$$= P(uz_1 - \bar{v}z_2, vz_1 + \bar{u}z_2) .$$

We can de-homogeneize by

$$p(z) = P(z,1) \quad (\text{conversely} : P(z_1, z_2) = z_2^n \, p(z_1/z_2)) .$$

Thus the above action on polynomials of degree $\leq n$ is given by

$$p = p(z) \longmapsto p_M = (vz+\bar{u})^n \, p\left(\frac{uz-\bar{v}}{vz+\bar{u}}\right)$$

$$p \in \Pi_n \quad \text{(space of complex polynomials of deg} \leq n \text{ in } z) .$$

In particular, the action of $M = -1_2$ is given by $p \longmapsto p_{-1} = (-1)^n p$ and we obtain a representation of

$$SU_2(\mathbb{C})/\{\pm 1\} \cong SO_3(\mathbb{R})$$

precisely when n is even : the corresponding dimension $\dim \Pi_n = n+1$ is odd. Let us take such an even integer $n = 2\ell$ and define the following diagram by commutativity

$$\begin{array}{ccc} SU_2(\mathbb{C}) & \xrightarrow{\tilde{D}_{n/2}} & Gl(\Pi_n) \\ & \searrow & \nearrow_{D_{n/2}} \\ & SO_3(\mathbb{R}) & \end{array} \qquad (\ell = \tfrac{1}{2}n \text{ integer}) .$$

We obtain two sequences of representations

$$\text{for } SU_2(\mathbb{C}) : \ \tilde{D}_o = \text{id.}, \tilde{D}_{\frac{1}{2}}, \tilde{D}_1, \ \ldots \ , \tilde{D}_{n/2}, \ldots$$

$$\dim \tilde{D}_{n/2} = n+1 \geq 1 ,$$

$$\text{for } SO_3(\mathbb{R}) : \ D_o = \text{id.}, D_1, D_2, \ldots , D_\ell, \ldots$$

$$\dim D_\ell = 2\ell + 1 \geq 1 .$$

(10.2) Theorem. The character of $\tilde{D}_{n/2}$ is given by

$$\tilde{\chi}_{n/2}(g) = U_n(\cos \tfrac{1}{2}\phi_g)$$

where $\frac{1}{2}\phi_g \in [0,\pi]$ is the angle of the rotation g and U_n is the nth Tchebycheff polynomial of the second kind. Similarly, for $\frac{1}{2}n = \ell$ integral, the character of D_ℓ is given by

$$\chi_\ell(g) = U_{2\ell}(\cos \tfrac{1}{2}\phi_g) \quad (\tfrac{1}{2}\phi_g \in [0,\tfrac{1}{2}\pi] \text{ angle of rotation } g).$$

Before we prove this statement, we remind the reader that the Tchebycheff polynomials of the first kind can be defined by

$$T_m(\cos\theta) = \cos m\theta .$$

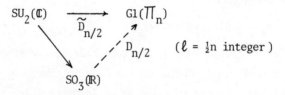

The sequence $T_o, T_1, \dots, T_n, \dots$ is orthogonal on $[-1,1]$ with respect to the density $(1-x^2)^{-\frac{1}{2}}$:

$$\int_{-1}^{+1} T_m(x)\, T_n(x)\,(1-x^2)^{-\frac{1}{2}}\,dx = 0 \quad \text{for } m \neq n .$$

These polynomials satisfy the normalization condition $T_m(1) = 1$. The <u>Tchebycheff polynomials of the</u> 2nd <u>kind</u> can similarly be defined by

$$U_m(\cos\theta) = \sin(m+1)\theta \,/\, \sin\theta \quad .$$

The sequence $U_o, U_1, \dots, U_n, \dots$ is orthogonal on $[-1,1]$ with respect to the density $(1-x^2)^{+\frac{1}{2}}$:

$$\int_{-1}^{+1} U_m(x)\, U_n(x)\, \sqrt{1-x^2}\,dx = 0 \quad \text{for } m \neq n .$$

They satisfy the normalization condition $U_m(1) = m+1$.

<u>Proof of the theorem</u>. Let us simply make the computations for $SU_2(\mathbb{C})$, taking the basis

$$1, z, z^2, \dots, z^n \quad \text{of } \Pi_n$$

consisting of simultaneous eigenvectors for the action of the matrices $\begin{pmatrix} e^{i\frac{1}{2}\varphi} & 0 \\ 0 & e^{-i\frac{1}{2}\varphi} \end{pmatrix}$. Such a matrix acts by

$$z^\nu \mapsto e^{-in\frac{1}{2}\varphi}\,(e^{i\varphi}z)^\nu = e^{i(2\nu-n)\frac{1}{2}\varphi}\, z^\nu \quad .$$

Thus the eigenvalues are

$$e^{-in\frac{1}{2}\varphi}, \dots, e^{in\frac{1}{2}\varphi}$$

and the corresponding trace is

$$\tilde{\chi}_{n/2}(g_\varphi) = \sum_{0\leqslant\nu\leqslant n} e^{i(\nu-\frac{1}{2}n)\varphi} = e^{-i\frac{1}{2}n\varphi}\,\frac{1-e^{i(n+1)\varphi}}{1-e^{i\varphi}} =$$

$$= \frac{e^{-i(n+1)\frac{1}{2}\varphi} - e^{i(n+1)\frac{1}{2}\varphi}}{e^{-i\frac{1}{2}\varphi} - e^{i\frac{1}{2}\varphi}} =$$

$$= \frac{\sin(n+1)\frac{1}{2}\varphi}{\sin\frac{1}{2}\varphi} = U_n(\cos\tfrac{1}{2}\varphi) \quad . \qquad \text{q.e.d.}$$

The reader has observed that in the case of $SO_3(\mathbb{R})$, only the polynomials $U_{2\ell}$ with even index appear. They form an orthogonal basis of the space of even polynomials on $[-1,1]$ (with the above density).

The representations $\widetilde{D}_{n/2}$ (of $\widetilde{G}$) and D_ℓ (of G) are irreducible for all $n, \ell \in \mathbb{N}$. To prove it, we shall use the criterion of (7.3). For this purpose, we have to know how to integrate central functions on G (or $\widetilde{G}$).

(10.3) Lemma. Take $G = SO_3(\mathbb{R})$ and $\varphi \in C_{inv}(G)$ and write

$$\varphi(g) = f(\cos \tfrac{1}{2}\phi_g) \qquad (\text{and } \tilde{f}(-x) = \tilde{f}(x) = f(x) \text{ for } -1 \leqslant x \leqslant 0).$$

Then we have

$$\int_G f(\cos \tfrac{1}{2}\phi_g)\, dg = \frac{4}{\pi}\int_0^1 f(x)\sqrt{1-x^2}\, dx = $$
$$= \frac{2}{\pi}\int_{-1}^1 \tilde{f}(x)\sqrt{1-x^2}\, dx$$

and

$$\int_G f(\cos \tfrac{1}{2}\phi_g)\, dg = c\int f(\cos \tfrac{1}{2}\phi)\sin^2 \tfrac{1}{2}\phi \; d(\tfrac{1}{2}\phi) \quad .$$

(The constant in this last formula is easily determined by taking $f = 1$, using the normalization of the Haar measure $\int_G dg = 1$.)

Proof. Let us make several changes of variables, with the purpose of ending up with

$$x = \cos \tfrac{1}{2}\phi \quad (= u_1 = \mathrm{Re}(u)) .$$

We start with

$$\rho = \cos \tfrac{1}{2}\theta \;, \quad \sigma = \tfrac{1}{2}(\varphi - \psi) \;, \quad \tau = \tfrac{1}{2}(\varphi + \psi) \quad .$$

The absolute value of the Jacobian of this transformation is $\frac{1}{4}\sin \frac{1}{2}\theta$. Thus we have

$$dg = (8\pi^2)^{-1} \sin\theta \, d\theta \, d\varphi \, d\psi =$$
$$= (8\pi^2 \cdot \tfrac{1}{4}\sin \tfrac{1}{2}\theta)^{-1} \sin\theta \, d\rho \, d\sigma \, d\tau = \pi^{-2}\rho \, d\rho \, d\sigma \, d\tau .$$

Then we put

$$\begin{cases} u_1 = \rho\cos\tau \quad (= \cos\tfrac{1}{2}\theta\cos\tfrac{1}{2}(\varphi+\psi) = \pm\cos\tfrac{1}{2}\phi) \\ u_2 = \rho\sin\tau \end{cases}$$

with Jacobian

$$\frac{\partial(u_1,u_2)}{\partial(\rho,\tau)} = \rho \;, \qquad \rho\, d\rho\, d\tau = du_1 du_2$$

(thus we have $u = u_1 + u_2 i = \rho e^{i\tau}$). This proves $dg = \pi^{-2} du_1 du_2 d\sigma$.

The domains of variation of these variables are indicated in the following pictures

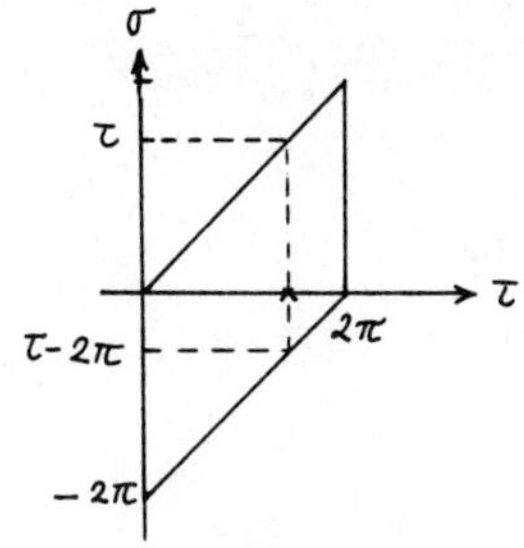

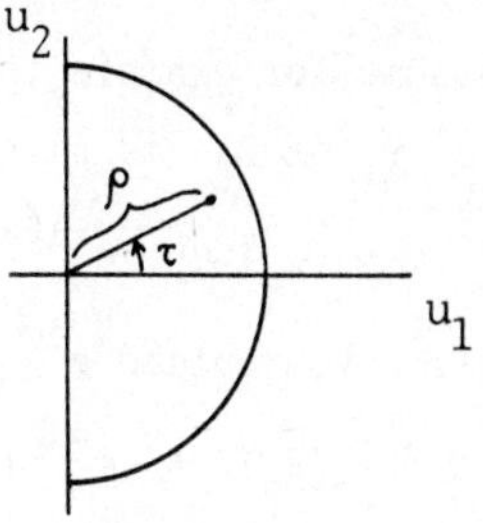

$\varphi, \psi \in [0,2\pi]$, τ fixed
$\sigma = \tau - \psi \in [\tau, \tau-2\pi]$

keeping u_1 and u_2 fixed
(i.e. ρ and τ fixed)
$\int_{\tau-2\pi}^{\tau} d\sigma = 2\pi$ (for all τ).

Integrating in u_2 (u_1 still fixed) we get

$$\int_{-\sqrt{1-u_1^2}}^{\sqrt{1-u_1^2}} du_2 = 2\sqrt{1-u_1^2} .$$

Finally,

$$\int_G f(\cos \tfrac{1}{2}\phi_g)\, dg = \pi^{-2}\cdot 2\pi\cdot 2 \int_0^1 du_1\, f(u_1)\sqrt{1-u_1^2} =$$

$$= \frac{4}{\pi}\int_0^1 f(x)\sqrt{1-x^2}\, dx = \frac{2}{\pi}\int_{-1}^1 \tilde{f}(x)\sqrt{1-x^2}\, dx ,$$

as claimed.

(10.4) Theorem. All representations $\tilde{D}_{n/2}$ and D_ℓ (of $\tilde{G}$, resp. G) are irreducible ($n, \ell \in \mathbb{N}$). More precisely, the dual of the compact group $\tilde{G}$ consists of the representations $\tilde{D}_{n/2}$ ($n \in \mathbb{N}$) and the dual of the compact group G consists of the representations D_ℓ ($\ell \in \mathbb{N}$).

Proof. The Tchebycheff polynomials of the second kind U_n ($n \in \mathbb{N}$) form a *total orthonormal* system on $[-1,1]$ with respect to the density $\frac{2}{\pi}\sqrt{1-x^2}\,dx$. The *even* polynomials $U_{2\ell}$ have the same property on $[0,1]$ with respect to the density $\frac{4}{\pi}\sqrt{1-x^2}\,dx$. Thus this theorem results from (7.3) and (10.3). q.e.d.

(10.5) Theorem (Clebsch-Gordan). For ℓ, $m \in \mathbb{N}$, the representation $D_\ell \otimes D_m$ of $SO_3(\mathbb{R})$ is equivalent to $D_{|\ell-m|} \oplus \ldots \oplus D_{\ell+m}$.

(Thus $D_\ell \otimes D_m$ is a sum of 2N + 1 inequivalent irreducible representations, the integer N being the minimum between ℓ and m.)

Proof. Assume for example that $m < \ell$. We have seen (10.2) that the character χ_ℓ of D_ℓ is given by

$$\chi_\ell(g_\varphi) = e^{-i\ell\varphi} + \dots + e^{i\ell\varphi} .$$

For simplicity, replace $e^{i\varphi}$ by t :

$$\chi_\ell(g_\varphi) = t^{-\ell}(1 + t + \dots + t^{2\ell}) = $$
$$= t^{-\ell}\,\frac{1 - t^{2\ell+1}}{1 - t} = \frac{t^{-\ell} - t^{\ell+1}}{1 - t} .$$

We have also seen that the character of a tensor product of representations is the usual product of characters (sec.7), hence

$$\chi_{D_\ell\otimes D_m}(g_\varphi) = \frac{t^{-\ell} - t^{\ell+1}}{1 - t} \sum_{|\mu| \leqslant m} t^\mu = \sum \frac{t^{-\ell+\mu} - t^{\ell+\mu+1}}{1 - t} .$$

But

$$\sum_{|\mu| \leqslant m} t^{-\ell+\mu} = \sum_{|\mu| \leqslant m} t^{-\ell-\mu} \qquad \text{(exchange } \mu \text{ and } -\mu\text{)}$$

so that we can still write

$$\chi_{D_\ell\otimes D_m}(g_\varphi) = \sum_{|\mu| \leqslant m} \frac{t^{-\ell-\mu} - t^{\ell+\mu+1}}{1 - t} =$$
$$= \chi_{\ell-m} + \chi_{\ell-m+1} + \dots + \chi_{\ell+m}$$

(when $m \leqslant \ell$). q.e.d.

EXERCISES

1. *For* $m \leq \ell \in \mathbb{N}$, *compute the sum of the dimensions of*

$$D_{\ell-m} \,,\ D_{\ell-m+1} \,,\ \dots \,,\ D_{\ell+m} \,.$$

(Either proceed directly without using (10.5) or use the fact that the dimension of a representation is the value at e of its character.)

2. *Determine the dual of the group* $U_2(\mathbb{C})$. *(Hint. Denote by* $Z = Z_2$ *the center of this group, so that* $Z \cap SU_2(\mathbb{C}) = \{\pm 1\}$. *Observe that any irreducible representation of* $U_2(\mathbb{C})$ *has an irreducible restriction to* $SU_2(\mathbb{C})$. *Identify finally the dual of* $U_2(\mathbb{C})$ *to the part of the product of the duals of Z and of* $SU_2(\mathbb{C})$ *consisting of representations agreeing on the matrix* -1_2 *.)*

3. *Let* π *be the canonical (3 dimensional) representation of* $G = SU_3(\mathbb{C})$.

a) *Show that* π *and* $\bar{\pi}$ *are inequivalent irreducible representations of G (use sec. 7).*

b) *Identify the representation* $\bar{\pi} \otimes \pi$ *with*

$$\begin{aligned} Ad : G &\longrightarrow GL(M) \,, \quad M = M_3(\mathbb{C}) = End(V) \text{ with } V = \mathbb{C}^3 \\ g &\longmapsto (A \longmapsto g A g^{-1}) \,. \end{aligned}$$

(Use the generators $\bar{u} \otimes v$ *of M with u and v in V.)*

c) *Give the decomposition of* $\bar{\pi} \otimes \pi$ *into irreducible components. (The subspaces* M_o *of zero trace matrices, and* M_1 *of scalar matrices are both invariant under the Ad representation.)*

d) *Let* ρ *be the eight dimensional irreducible representation of G. Show that the restriction of* ρ *to the diagonal subgroup T of G contains the identity representation of T with multiplicity 2. (This representation* ρ *is used in physics to interpret the symmetry structure of octets of baryons and of mesons. In particular, the meson* η *- companion of the meson* π^0 *- was invented before it was discovered simply to account for the above multiplicity 2. The three-dimensional representations* π *and* $\bar{\pi}$ *correspond to triplets of quarks...)*

4. *Compute the following coefficient (up to a multiplicative constant)*

$$(z^{\ell} \mid \tilde{D}_{\ell}(a_{\theta})\, z^{\ell}) \qquad (a_{\theta} = \text{image of } \begin{pmatrix} \cos\theta & \sin\theta \\ -\sin\theta & \cos\theta \end{pmatrix} \text{ in } SO_3(\mathbb{R}) \ )$$

of the representation $\tilde{D}_{\ell}$ *of* $SO_3(\mathbb{R})$ *in the space of polynomials in z of degree* $\leqslant 2\ell$. *(Observe that since the basis* $1, z, \ldots, z^{2\ell}$ *is an orthogonal basis, this coefficient is proportional to the coefficient of* z^{ℓ} *in the expansion of*

$$\tilde{D}_{\ell}(a_{\theta})\, z^{\ell} = (z\sin\theta + \cos\theta)^{\ell}\,(z\cos\theta - \sin\theta)^{\ell} \ .)$$

In particular, show that this coefficient is proportional to the Legendre polynomial $P_{\ell}(\cos 2\theta)$. *(Recall that*

$$2^{\ell}\ell!\; P_{\ell}(\zeta) = \frac{d^{\ell}}{d\zeta^{\ell}}(\zeta^2 - 1)^{\ell} \qquad .)$$

More generally, show that the coefficient $(z^{\ell+m} \mid \tilde{D}_{\ell}(a_{\theta})\, z^{\ell})$ *is proportional to the associated Legendre function*

$$P_{\ell}^{m}(\cos 2\theta) \qquad\qquad (-\ell \leqslant m \leqslant \ell)$$

defined by

$$P_{\ell}^{m}(\zeta) = (1 - \zeta^2)^{\frac{1}{2}|m|}\, \frac{d^{|m|}}{d\zeta^{|m|}}\, P_{\ell}(\zeta) \ .$$

Using Ex.5 of sec.7 (part c), give an integral relation between Legendre polynomials and Tchebycheff polynomials of the second kind.

Part II :

Representations of locally compact groups

11 GROUPS WITH FEW FINITE-DIMENSIONAL REPRESENTATIONS

In this section, we show that for non-compact groups, the consideration of finite-dimensional representations is not sufficient.

Theorem. Let G be a connected solvable group. Then any finite dimensional irreducible representation of G has dimension 1.

Proof. Denote by $G' = \overline{[G,G]}$ the closure of the subgroup generated by the commutators of G, and by induction $G^{(i)} = (G^{(i-1)})'$. Thus $G^{(i)}$ is a closed normal subgroup of $G^{(i-1)}$ and since G is solvable, $G^{(h)} = \{e\}$ for some integer $h \in \mathbb{N}$ that we take minimal

$$G \supset G' \supset G'' \supset \quad \dots \quad \supset G^{(h)} = \{e\} .$$

Let us call this minimal integer h such that $G^{(h)} = \{e\}$, the *length* of G. We shall prove the theorem by induction on this length h. For h = 1 (i.e. G commutative), the theorem follows from Schur's lemma. Let now $h > 1$ and assume that the theorem has already been proved for solvable groups of length $\leqslant h-1$. Take any finite-dimensional irreducible representation (π, V) of G and decompose the restriction of π to G' into irreducible components (by induction assumption, these irreducible components will have dimension 1). For $\lambda \in \mathrm{Hom}(G',\mathbb{C}^{\times})$, we put

$$V_{\lambda} = \{v \in V : \pi(s)\, v = \lambda(s)\, v\} .$$

The non-zero subspaces V_{λ} are linearly independent : take any linear dependence relation

$$\sum_{1 \leqslant i \leqslant k} a_i v_i = 0 \qquad\qquad (v_i \in V_{\lambda_i})$$

with k *minimal*. Then, obviously $\sum \lambda_1(s)\, a_i v_i = 0$ and also

$$0 = \pi(s) \sum a_i v_i = \sum \lambda_i(s)\, a_i v_i$$

whence by subtraction

$$\sum_{2 \leqslant i \leqslant k} (\lambda_1(s) - \lambda_i(s))\, a_i v_i = 0$$

contradicting the minimality of k. In particular, the set of $\lambda \in \mathrm{Hom}(G',\mathbb{C}^{\times})$

such that $V_\lambda \neq \{0\}$ is *finite*. On the other hand, the group G acts on this set of characters (by conjugation) : if $v \in V_\lambda$

$$\pi(s)\pi(g)\,v = \pi(g)\pi(g^{-1}sg)\,v = \pi(g)\lambda(g^{-1}sg)\,v \qquad (s,g \in G),$$

so that with $\lambda_g(s) = \lambda(g^{-1}sg)$ we have

$$\pi(g)\,V_\lambda \subset V_{\lambda_g} \qquad (g \in G).$$

The action of G on this finite set of characters is continuous, and G being connected, it must be trivial : $\lambda_g = \lambda$ for all $g \in G$. Take now any $V_\lambda \neq \{0\}$. The preceding argument shows that V_λ is invariant under the action of the whole group G, hence $V_\lambda = V$ by irreducibility. Thus G' acts by scalar multiplications in V

$$\pi|_{G'} = \lambda \cdot \mathrm{id}.$$

Let us now take an element $s \in G$ and an eigenvector v of $\pi(s)$, say

$$\pi(s)\,v = \mu v \qquad (v \neq 0,\ \mu \neq 0\ ).$$

From the relation

$$\pi(s)\,\pi(g) = \pi(g)\pi(s)\lambda(s^{-1}g^{-1}sg)$$

follows that $\pi(g)\,v$ is also an eigenvector of $\pi(s)$ corresponding to the eigenvalue

$$\mu\,\lambda(s^{-1}g^{-1}sg) = \mu\lambda^{(g)} \quad .$$

But the set of eigenvalues of $\pi(s)$ is finite and the action of G on this set is continuous. As before, G being connected, $\mu\lambda^{(g)}$ must be independent of g

$$\mu\,\lambda^{(g)} = \mu\,\lambda^{(e)} = \mu \quad .$$

This proves that the μ-eigenspace of $\pi(s)$ is G-invariant (and $\neq \{0\}$) hence equal to the whole space V by irreducibility : $\pi(s) = \mu\cdot\mathrm{id}$. . This is true for any $s \in G$: G acts scalarly in V and by irreducibility, $\dim(V) = 1$. q.e.d.

Corollary 1. Let G be as in the theorem and denote by G_o the intersection of the kernels of all finite-dimensional irreducible representations of G. Then G_o contains the commutator subgroup G'.

Corollary 2. Let G be the solvable subgroup of $Sl_2(\mathbb{R})$ consisting of the matrices

$$\begin{pmatrix} t & u \\ 0 & 1/t \end{pmatrix} \qquad (t > 0, u \in \mathbb{R})$$

and U its subgroup defined by t = 1. Then, in any finite-dimensional irreducible representation of G, U acts trivially.

Proof. Just observe that G is connected and G' = U.

Corollary 3. Any finite-dimensional unitary representation of the group $Sl_2(\mathbb{R})$ is trivial.

Proof. Let G be the connected solvable subgroup of $Sl_2(\mathbb{R})$ defined in the corollary 2. By (2.1), the restriction to G of any finite-dimensional unitary representation can be decomposed as the sum of irreducible ones. By Cor.2 above, the matrices

$$\begin{pmatrix} 1 & u \\ 0 & 1 \end{pmatrix}, \quad \begin{pmatrix} 1 & 0 \\ u & 1 \end{pmatrix} = \begin{pmatrix} 0 & -1 \\ 1 & 0 \end{pmatrix}\begin{pmatrix} 1 & -u \\ 0 & 1 \end{pmatrix}\begin{pmatrix} 0 & -1 \\ 1 & 0 \end{pmatrix}^{-1}$$

act trivially in all these irreducible components, hence act trivially. Now, if a matrix

$$\begin{pmatrix} a & b \\ c & d \end{pmatrix} \in Sl_2(\mathbb{R})$$

has a coefficient $c \neq 0$, it can be written

$$\begin{pmatrix} a & b \\ c & d \end{pmatrix} = \begin{pmatrix} 1 & (a-1)/c \\ 0 & 1 \end{pmatrix}\begin{pmatrix} 1 & 0 \\ c & 1 \end{pmatrix}\begin{pmatrix} 1 & (d-1)/c \\ 0 & 1 \end{pmatrix}$$

hence acts also trivially. Finally, if c = 0

$$\begin{pmatrix} a & b \\ 0 & d \end{pmatrix} = \begin{pmatrix} -b & a \\ -d & 0 \end{pmatrix}\begin{pmatrix} 0 & -1 \\ 1 & 0 \end{pmatrix}$$

must also act trivially as we have just seen (product of two matrices with $c \neq 0$).

Corollary 4. The finite-dimensional unitary irreducible representations of the Heisenberg group G consisting of the matrices

$$\begin{pmatrix} 1 & a & c \\ 0 & 1 & b \\ 0 & 0 & 1 \end{pmatrix} \qquad (a, b, c \in \mathbb{R})$$

have dimension 1 and correspond to the characters of the abelianized group $G_{ab} = G/G' \cong \mathbb{R}^2$.

Proof. In our case, the derived subgroup G' consists of the matrices with a = b = 0 (and coïncides with the center of G). Thus G' acts trivially in any finite-dimensional irreducible representation. In particular, G' will also act trivially in any finite-dimensional completely reducible representation. q.e.d.

It is possible to show that for any non-trivial character $\chi : Z = G' \rightarrow \mathbb{C}$ of the center of the Heisenberg group G, there is one and - up to equivalence - only one (infinite dimensional) representation π of G in a Hilbert space inducing the character χ on Z. Moreover, π is equivalent to a unitary representation precisely when its central character χ is unitary (Ex. 2 at the end of this sec.).

The above corollary 3 admits the following generalization.

Theorem. Let G be a connected semi-simple real Lie group having no invariant compact subgroup $\neq \{e\}$. Then the only finite-dimensional unitary irreducible representation of G is the trivial representation (in dimension 1).

Sketch of proof. It is enough to prove this theorem when G is simple (the center Z of G is then a discrete subgroup and G/Z is simple in the abstract algebraic sense). We use the following properties :

a) there exists a one-parameter subgroup $\mathbb{R} \cong H \subset G$ and elements $g_n \in G$ such that $g_n h g_n^{-1} \rightarrow e$ for all $h \in H$,

b) if H is a non-trivial one-parameter subgroup, the union of all conjugates of H contains a neighbourhood of e in G.

From these, we infer that for any finite-dimensional representation (π, V) of G

$$\pi(g_n)\,\pi(h)\,\pi(g_n)^{-1} \rightarrow 1_V \qquad (h \in H)$$

whence

$$d = \dim V = \mathrm{Tr}\, 1_V = \lim \mathrm{Tr}\, \pi(g_n)\,\pi(h)\,\pi(g_n)^{-1} = \mathrm{Tr}\, \pi(h) .$$

Fix $h \in H$ and put $\pi(h) = A = R + iS$ with $R = \frac{1}{2}(A + A^*)$ and $S = \frac{1}{2i}(A - A^*)$ both hermitian. Then $d = \mathrm{Tr}\, A = \mathrm{Tr}\, R + i\,\mathrm{Tr}\, S$ so that

$$\mathrm{Tr}\, S = 0 , \quad \mathrm{Tr}\, R = d \quad \text{(a priori, these traces are } \textit{real}\text{)}.$$

But π is unitary :

$$\|A\| = \|\pi(h)\| = 1 \implies \|R\| = \|\tfrac{1}{2}(A+A^*)\| \leqslant 1 .$$

Since R is a hermitian operator, we infer that

$$R \leqslant 1 \quad \text{and} \quad 1 - R \geqslant 0 ,$$

so that

$$\mathrm{Tr}(1 - R) = \mathrm{Tr}\, 1 - \mathrm{Tr}\, R = d - d = 0 .$$

But the positive operator 1 - R can have zero trace only if it is zero

$$R = 1 \quad , \quad A = 1 + iS \quad .$$

But A is unitary :

$$1 = AA^* = (1 + iS)(1 - iS) = 1 + S^2$$

$$\implies S^2 = 0 \implies S = 0 \implies A = 1 .$$

This proves that $\pi(h) = 1$ for $h \in H$. Since $\mathrm{Ker}\,\pi$ is a normal subgroup of G, it contains all conjugates of H, and by b) above, contains a neighbourhood of $e \in G$. But G is connected, hence generated by any neighbourhood of its neutral element. q.e.d.

EXERCISES

1. *Let G be a compact solvable group. Show that the neutral connected component of G is commutative (this neutral connected component is solvable : one can use the first theorem of this section and (4.1)). In particular, any connected compact solvable group is commutative. But there are some non-commutative finite solvable groups ! Also show that if p is any prime integer, the subgroup of* $Gl_2(\mathbb{Q}_p)$ *consisting of the matrices*

$$\begin{pmatrix} a & b \\ 0 & 1 \end{pmatrix} \text{ with } a \in \mathbb{Z}_p^{\times} \text{ and } b \in \mathbb{Z}_p$$

is solvable (and compact !).

2. *Let G be the Heisenberg group and* $\chi : \mathbb{R} \longrightarrow \mathbb{C}^{\times}$ *a character. We consider the space* V_χ *of functions f on G satisfying*

$$f\begin{pmatrix} 1 & x & z \\ 0 & 1 & y \\ 0 & 0 & 1 \end{pmatrix} = \varphi(x)\,\chi(z) \quad \text{for some } \varphi \in L^2(\mathbb{R}).$$

With the scalar product of the space $L^2(\mathbb{R})$, V_χ *is a Hilbert space.*

a) *Show that the right regular representation of G acts on* V_χ *(it leaves* V_χ *invariant).*

b) *The action of the center of G on* V_χ *is given by the character* χ.

c) *The representation* V_χ *is unitary precisely when* χ *is.*

d) *Assume* χ *unitary and* $\chi \neq$ *id. so that the characters*

$$\chi_b : x \longmapsto \chi(bx) \qquad (b \in \mathbb{R})$$

represent all unitary characters of $\mathbb{R}$. *Irreducibility of* V_χ *is then equivalent to irreducibility of* $L^2(\mathbb{R})$ *under translations and multiplication by unitary characters.*

12 INVARIANT MEASURES ON LOCALLY COMPACT GROUPS AND HOMOGENEOUS SPACES

Let G be a locally compact group and $C_c(G)$ the vector space of (complex) continuous functions on G which vanish outside some compact subset of G. A Haar measure on G is a positive continuous linear form

$$m : C_c(G) \to \mathbb{C}, \quad f \mapsto m(f) = \int_G f(x)\,dx$$

invariant under *left* translations

$$m({}_sf) = m(f) \qquad (\text{for } f \in C_c(G),\ s \in G,\ {}_sf(x) = f(s^{-1}x)\ ).$$

Continuity refers to the following property: for each compact subset K of G, there exists a constant C_K such that

$$|m(f)| \leqslant C_K \operatorname{Sup}_x |f(x)| \text{ for all } f \in C_c(G) \text{ vanishing outside K.}$$

Positivity simply means $m \neq 0$ and

$$f \geqslant 0\ (f \in C_c(G)\ ) \implies m(f) \geqslant 0\ .$$

The *existence* of a Haar measure on any locally compact group is a standard (technical) result that we are going to take for granted (cf. Weil, 1953 p.34 or Bourbaki 1963, Int. Chap.7 p.13). *Uniqueness* up to a positive multiplicative constant is however easily proved as follows. Let m and m' be two Haar measure, and define the measure n by

$$n(f) = m'(f^{\vee}) \qquad (f \in C_c(G)\ ).$$

Thus the measure n is invariant under *right* translations. We show first that n has a density with respect to m : for f and $g \in C_c(G)$

$$m(f)\,n(g) = m(f)\int g(ts)\,dn(t) = \int dm(s)\,f(s)\int dn(t)\,g(ts) =$$

$$= \int dn(t)\int dm(s)\,f(s)\,g(ts) = \int dn(t)\int dm(x)\,f(t^{-1}x)\,g(x) =$$

$$= \int dm(x)\,g(x)\int dn(t)\,f(t^{-1}x)\quad .$$

From this follows that $n = w\,m$ with a density function

$$w(x) = \int f(t^{-1}x)\,dn(t)\,/m(f)$$

independent of f provided $m(f) \neq 0$ (e.g. $f \geqslant 0$ and $0 \neq f \in C_c(G)$).

Observe that this density function w is continuous (it is given by an integral depending continuously in the parameter x) and the support of n is the whole group G, so that w is the unique continuous function with $n = wm$. Take now $x = e$ in the definition of w

$$w(e)\, m(f) \;=\; n(f^{\vee}) \;=\; m'(f) \quad \text{for all } f \geqslant 0,\ 0 \neq f \in C_c(G).$$

This proves $m' = w(e)\, m$ proportional to m as claimed.

From the relative uniqueness of Haar measures follows that if $\alpha : G \to G$ is a (continuous) automorphism of G,

$$f \mapsto m(f^{\alpha}) \quad \text{where} \quad f^{\alpha}(x) = f(\alpha^{-1}(x)) \qquad (f \in C_c(G))$$

is proportional to m, say

$$m(f^{\alpha}) = \mathrm{mod}_G(\alpha)\, m(f) \quad .$$

With the integral notation, this relation can be written

$$\int_G f(\alpha^{-1}(x))\, dx = \mathrm{mod}_G(\alpha) \int_G f(x)\, dx \qquad (dx = dm(x)).$$

In particular when $\alpha = \mathrm{Int}(s)$ is the inner automorphism produced by an element $s \in G$, we put $\Delta_G(s) = \mathrm{mod}_G(\mathrm{Int}(s))$

$$\Delta_G : G \to \mathbb{R}_+^{\times} \quad , \qquad \underline{\text{modular function of}}\ G\ .$$

By definition, we have thus

(12.1) $$\int_G f(sxs^{-1})\, dx = \int_G f(xs^{-1})\, dx = \Delta_G(s) \int_G f(x)\, dx \ .$$

Extending the Haar measure m to Borel or measurable subsets of G, the preceding relations can be written in the perhaps more natural form

(12.2) $$\begin{cases} m(\alpha(A)) = \mathrm{mod}_G(\alpha)\, m(A) & , \\ m(s^{-1}As) = m(As) = \Delta_G(s)\, m(A) & . \end{cases}$$

The modular function $\Delta_G : G \to \mathbb{R}_+^{\times}$ is obviously a homomorphism. It is continuous since by (12.1) we can write it

$$\Delta_G(s) = \int_G f(xs^{-1})\, dx \Big/ \int_G f(x)\, dx$$

for any $f \in C_c(G)$ with $m(f) \neq 0$. In fact, since obviously $\Delta_G(s) = 1$ for s in the center Z of G, the modular function can be regarded as a function over G/Z:

(12.3) $$\Delta_G = 1 \quad \underline{\text{when}}\ G/Z\ \underline{\text{is a simple group}}\ ,\ \underline{\text{or}}\ G/Z\ \underline{\text{compact}}\ .$$

(12.4) Lemma. Let G be a locally compact group, $dx = dm(x)$ a Haar measure on G. Then with $\Delta = \Delta_G$

$$m(f^{\vee}) = \int_G f(x^{-1})\,dx = \int_G \frac{f(x)}{\Delta(x)}\,dx = m(f/\Delta)$$

for $f \in C_c(G)$ (or $f \in L^1(G)\ldots$). More simply $d(x^{-1}) = \Delta(x)^{-1}\,dx$.

Proof. One checks immediately that the two measures $d(x^{-1})$ and $\Delta(x)^{-1}\,dx$ are invariant under right translations, hence proportional

$$d(x^{-1}) = c\,\Delta(x)^{-1}\,dx \quad .$$

Replacing x by x^{-1} in the preceding equality, we have $dx = c\,\Delta(x)\,d(x^{-1})$ and substituting the expression for $d(x^{-1})$

$$dx = c\,\Delta(x)\;c\,\Delta(x)^{-1}\,dx = c^2\,dx \quad .$$

Since all these measures are positive, $c > 0$ and $c = 1$ q.e.d.

(12.5) Proposition. Let G be a locally compact group and H a closed normal subgroup of G. Then the modular function Δ_H of H is the restriction to H of the modular function Δ_G of G.

Proof. Since H is closed, G/H is a Hausdorff space. Moreover, since the canonical projection $p : G \to G/H$ is an open map by the definition of the quotient topology, for any compact neighbourhood V of e in G, p(V) is a compact neighbourhood of $\dot{e}$ in G/H. In particular, G/H is a locally compact group. Let $\dot{m}$ denote a Haar measure of G/H. Then the linear form

$$f \mapsto \int_{G/H} d\dot{m}(\dot{x}) \int_H f(xh)\,dh \qquad (f \in C_c(G))$$

is a Haar measure m on G. Replacing f by a right translate f_s (for $s \in G$ we put $f_s(x) = f(xs)$) we obtain immediately when $s \in H$

$$\Delta_G(s)^{-1} = \Delta_H(s)^{-1} \quad .$$

q.e.d.

We say that a locally compact group G is unimodular when its modular function is trivial: $\Delta_G = 1$. Unimodular groups are those having a *left and right* invariant (Haar) measure.

(12.6) Corollary. Let G be a locally compact group. Then the kernel of the modular function Δ_G is the largest closed invariant unimodular subgroup of G.

HOMOGENEOUS SPACES

Let H be a closed subgroup of the locally compact group G. We are going to investigate when there exists an invariant measure on the homogeneous space G/H (G acts by left translations on this space). If dx and dh are Haar measures on G and H respectively, any invariant measure m (or $dm(\dot{x})$) on G/H should satisfy

$$(*) \qquad \int_G f(x)\,dx = \int_{G/H} dm(\dot{x}) \int_H f(xh)\,dh \qquad (f \in C_c(G)) .$$

Applying this equality to a right translate of f by an element $s \in H$ gives the condition $\Delta_G(s) = \Delta_H(s)$. Thus, an invariant measure on G/H can exist only if

$$\Delta_H = \Delta_G\Big|_H \quad .$$

We are going to show that this condition is sufficient. Quite generally, let us define a function f^H on G/H when $f \in C_c(G)$ by

$$f^H(\dot{x}) = \int_H f(xh)\,dh \quad .$$

Thus, (*) can be rewritten

$$\int_G f(x)\,dx = \int_{G/H} dm(\dot{x})\, f^H(\dot{x}) \quad .$$

In particular, we see that if an invariant measure m on G/H exists,

$$(**) \qquad f^H = 0 \implies \int_G f(x)\,dx = 0 \qquad (f \in C_c(G)).$$

This is the crucial consequence of (*). Indeed, if (**) is satisfied, the value of m on the *special functions* $\varphi = f^H \in C_c(G/H)$ (for some $f \in C_c(G)$) is given by

$$m(\varphi) = m(f^H) = \int_G f(x)\,dx \quad .$$

independently from the choice of f :

$$f^H = g^H \implies (f-g)^H = 0$$

$$\underset{(*)}{\implies} \int_G (f-g)\,dx = 0 \implies \int_G f\,dx = \int_G g\,dx \quad .$$

According to these preliminary observations, the strategy is as follows

a) prove that <u>all</u> $\varphi \in C_c(G/H)$ can be written $\varphi = f^H$ (for some $f \in C_c(G)$) ,

b) show that $\Delta_H = \Delta_G \big|_H \implies$ (*) holds .

For reference, we formulate these two steps as follows.

(12.7) Lemma. The linear mapping $C_c(G) \to C_c(G/H)$, $f \mapsto f^H$ is surjective.

Proof. First of all, letting $p : G \to G/H$ denote the canonical projection, we show that every compact subset A of G/H is the image A = p(K) of some compact subset K of G. Indeed, when U is an open relatively compact set of G, p(U) is also open and relatively compact in G/H. These sets cover A so that we can find a finite number of open relatively compact subsets U_i of G for which the union of the $p(U_i)$ cover A. Thus, we can take for K

$$\overset{-1}{p}(A) \cap \bigcup \overline{U}_i \quad .$$

Then, it is enough to show that a positive function $\varphi \in C_c(G/H)$ with compact support A = p(K) in G/H is of the form f^H for some positive continuous function f with support in $K \subset G$. But there exists a

$$\theta \in C_c(G) \ , \quad \theta > 0 \text{ on } K \text{ (hence } \theta > 0 \text{ in a neighbourhood of } K).$$

The obvious formula

$$(\varphi \cdot \theta)^H = \varphi \cdot \theta^H$$

shows immediately that $\varphi = (\theta \cdot \varphi/\theta^H)^H = f^H$ with

$$f = \theta \cdot \varphi / \theta^H \quad .$$

Since θ has compact support, f also has compact support and since $\theta^H > 0$ on the support of φ, φ/θ^H is well-defined and continuous. q.e.d.

(12.8) Proposition. Let H be a closed subgroup of the locally compact group G. An invariant measure exists on G/H if and only if $\Delta_H = \Delta_G \big|_H$. Such an invariant measure m on G/H is characterized by

$$\int_G f(x)\,dx = \int_{G/H} dm(\dot{x}) \int_H f(xh)\,dh = \int_{G/H} f^H(\dot{x})\,dm(\dot{x})$$

for all $f \in C_c(G)$.

Proof. It only remains to show that the assumption on the modular function of H implies that (*) holds. But when f and $g \in C_c(G)$, Fubini's theorem (for continuous functions on compact spaces) gives

$$\int_G f(x)\, g^H(x)\, dx \;=\; \int_G dx\, f(x) \int_H dh\, g(xh) \;=$$

$$= \int dh \int dx\, f(x)\, g(xh) \;=\; \int dh \int dx\, f(xh^{-1})\, g(x)\, \Delta_G(h)^{-1} \;.$$

By assumption, and since $d(h^{-1}) = \Delta_H(h)^{-1}\, dh$ (12.4), we can still write the preceding integral as

$$\int dx \int f(xh^{-1})\, g(x)\, \Delta_H(h)^{-1}\, dh \;=\; \int dx \int f(xh)\, g(x)\, dh \;=$$

$$= \int_G f^H(x)\, g(x)\, dx \;.$$

Consequently, $f^H = 0$ implies

$$\int_G f(x)\, g^H(x)\, dx \;=\; 0 \quad \text{for all} \quad g \in C_c(G) \;.$$

But by the preceding lemma (12.7), there exists a function $g \in C_c(G)$ for which $g^H(x) = 1$ for all $x \in \mathrm{Supp}(f)$. Thus

$$\int_G f(x)\, dx \;=\; 0$$

as was to be shown. q.e.d.

One should observe that when H is a compact subgroup of G, the condition of the proposition (12.8) is automatically satisfied. In this case, $p : G \longrightarrow G/H$ is a proper map and its transpose

$$^t p \;:\; C_c(G/H) \;\longrightarrow\; C_c(G) \quad \text{(composition with p)}$$

allows one to define the invariant measure m on G/H simply by composition of the Haar measure of G with $^t p$: m is then the *image of the Haar measure* of G by the proper map p (cf. proof of (3.1) and (8.7)).

When a quotient G/H has no invariant measure (cf. examples below), it may still be useful to consider *relatively invariant* measures on this homogeneous space. By definition, these measures are those which are multiplied by some constants under translations. More generally, *quasi-invariant* measures on G/H are those which are multiplied by functions under translations. According to the Radon-Nikodym theorem, the quasi-invariant measures are characterized by the fact that their family of negligible sets is invariant under translations.

EXAMPLES

1) The Heisenberg group consisting of matrices

$$\begin{pmatrix} 1 & x & z \\ 0 & 1 & y \\ 0 & 0 & 1 \end{pmatrix} \qquad (x, y, z \in \mathbb{R})$$

is unimodular. Indeed, the measure $dx\,dy\,dz$ is both left and right invariant.

2) Let K be a locally compact (non-discrete) field and dx a Haar measure on the additive group of K. For any $a \in K^\times$, $x \mapsto ax$ is an automorphism of this additive group and thus

$$d(ax) = \mathrm{mod}_K(a)\,dx = |a|_K\,dx$$

for some positive constant $|a|_K$. Moreover, $a \mapsto |a|_K$ is a continuous homomorphism $K^\times \to \mathbb{R}_+^\times$. On the multiplicative group $K^\times$, $d^\times x = |x|_K^{-1}\,dx$ is a Haar measure.

3) Let G denote the subgroup of upper triangular matrices of $Sl_2(\mathbb{R})$. The identity

$$\begin{pmatrix} a & b \\ 0 & a^{-1} \end{pmatrix}\begin{pmatrix} x & y \\ 0 & x^{-1} \end{pmatrix} = \begin{pmatrix} ax & ay+bx^{-1} \\ 0 & a^{-1}x^{-1} \end{pmatrix}$$

shows that

$$x^{-2}\,dx\,dy$$

is a Haar measure on G (observe that $d(ay + bx^{-1}) = d(ay) = |a|dy$).

On the other hand,

$$\begin{pmatrix} x & y \\ 0 & x^{-1} \end{pmatrix}\begin{pmatrix} a & b \\ 0 & a^{-1} \end{pmatrix} = \begin{pmatrix} ax & a^{-1}y + bx \\ 0 & a^{-1}x^{-1} \end{pmatrix}$$

shows that

$$\Delta\begin{pmatrix} a & b \\ 0 & a^{-1} \end{pmatrix} = 1/a^2 .$$

In particular, this solvable group is not unimodular. The reader will check that a Haar measure on the group G_1 consisting of matrices

$$\begin{pmatrix} x & y \\ 0 & 1 \end{pmatrix} \qquad (x \in \mathbb{R}^\times, y \in \mathbb{R})$$

is given by $x^{-2}\,dx\,dy$. Here, $\Delta\begin{pmatrix} a & b \\ 0 & 1 \end{pmatrix} = |a|^{-1}$ and $|x|^{-1}\,dx\,dy = d^\times x\,dy$ is a right invariant measure. Similar considerations hold when the field $\mathbb{R}$ is replaced by any locally compact field K.

4) Let K be a locally compact (non-discrete) field. We are going to determine a Haar measure on the group $Gl_n(K)$. Write $g = (g^i_j)$ for an element of this group. The product measure

$$\bigotimes_{i,j} dg^i_j$$

is not invariant, but it is easy to compute its transforms under translations. Let $V = K^n$, $\mathfrak{A} = \mathrm{End}(V) = V^\vee \otimes V$, (e_i) the canonical basis of V and (e^i) its dual (canonical basis of $V^\vee$). We also denote by $P_i = e^i \otimes e_i$ the projectors on the lines Ke_i generated by the basis vectors. Then

$$\begin{aligned} \mathfrak{A} &= \bigoplus_i \mathfrak{A} P_i \xrightarrow{\sim} \bigoplus_i V \\ x &= \sum e^i \otimes x_i \mapsto (x_i) \\ & \qquad T \mapsto (Te_i) \end{aligned}$$

is a left $\mathfrak{A}$-isomorphism :

$$\begin{aligned} \left(A \cdot \sum e^i \otimes x_i\right)(x) &= A \sum e^i(x)\, x_i = \\ &= \sum e^i(x)\, g(x_i) = \sum e^i \otimes g(x_i)\ (x) . \end{aligned}$$

From this follows that left multiplication by a $g \in Gl_n(K)$ in $\mathfrak{A}$ has a determinant equal to $(\det g)^n$ and consequently

$$dg = |\det g|_K^{-n} \bigotimes_{i,j} dg^i_j$$

is a Haar measure on $Gl_n(K)$. Similar computations hold with respect to right translations, hence $Gl_n(K)$ is a unimodular group. Its invariant subgroup $Sl_n(K)$ is thus also unimodular.

5) Take now in particular the unimodular group $G = Sl_2(\mathbb{R})$. It is useful to let G act on the upper half-plane $y = \mathrm{Im}(z) > 0$, $z = x + iy \in \mathbb{C}$ by fractional linear transformations

$$g = \begin{pmatrix} a & b \\ c & d \end{pmatrix} \text{ acting by } g\cdot z = \frac{az+b}{cz+d} .$$

This action is transitive (the subgroup with c = 0 already acts transitively) and the stabilizer of the point i is a compact subgroup K :

$$\frac{ai+b}{ci+d} = i \iff ai + b = id - c \iff a = d \ \&\ b = -c .$$

Thus $K = SO_2(\mathbb{R}) \subset G = Sl_2(\mathbb{R})$ and $g \mapsto g\cdot i$ gives an isomorphism of the homogeneous space G/K with the upper half-plane . We first determine an invariant measure on the upper half-plane. Let

$$g = \begin{pmatrix} a & b \\ c & d \end{pmatrix}, \quad w = g\cdot z = \frac{a z + b}{c z + d},$$

so that

$$w = |c z + d|^{-2} (a z + b)(c \bar{z} + d) =$$

$$= |c z + d|^{-2} (\dots + i(ad - bc)),$$

$$\operatorname{Im}(w) = |c z + d|^{-2} \operatorname{Im}(z)$$

and

$$dw = (cz + d)^{-2} (a(cz + d) - c(az + b)) \, dz =$$

$$= (cz + d)^{-2} \, dz .$$

In particular

$$d\bar{w} \wedge dw = |cz + d|^{-4} \, d\bar{z} \wedge dz ,$$
$$\operatorname{Im}(w)^2 = |cz + d|^{-4} \operatorname{Im}(z)^2$$

and

$$\operatorname{Im}(z)^{-2} \, d\bar{z} \wedge dz$$

is invariant under fractional linear transformations. Since $d\bar{z} \wedge dz = 2i(dx \wedge dy)$, a *positive* invariant measure on the upper half-plane is given by

$$y^{-2} \, | dx \wedge dy | = y^{-2} \, dx \, dy .$$

The formula

$$\int_G f(g) \, dg = \int_{G/K} dm(\dot{g}) \int_K f(gk) \, dk \qquad (f \in C_c(G))$$

shows that with the identification $gK \mapsto g\cdot i = x + iy$ we can write

$$\int_G f(g) \, dg = \int_{y = \operatorname{Im} z > 0} y^{-2} \, dx \, dy \int_{SO_2(\mathbb{R})} f(gk) \, dk .$$

Of course, the Haar measure of the circle group $K = SO_2(\mathbb{R})$ is

$$\frac{1}{2\pi} \, d\theta \quad \text{if} \quad g = g_\theta = \begin{pmatrix} \cos\theta & \sin\theta \\ -\sin\theta & \cos\theta \end{pmatrix} \qquad (\theta \in [0, 2\pi[).$$

Observe however, that if B denotes the subgroup of upper triangular matrices in $Sl_2(\mathbb{R})$, the modular function $\Delta_B \neq 1$ does not coïncide with the restriction of the modular function Δ_G (example 3 above) and thus there is no invariant measure on the homogeneous space G/B .

EXERCISES

1. Let G be a locally compact group and H a closed subgroup of G. Assume that there is a relatively invariant measure m or $dm(\dot{x})$ on G/H. By definition, there are identities

$$\int_{G/H} \varphi(s^{-1}\dot{x})\, dm(\dot{x}) = \chi(s) \int_{G/H} \varphi(\dot{x})\, dm(\dot{x}) \qquad (\varphi \in C_c(G/H)).$$

a) Show that χ is a character (i.e. a continuous homomorphism) $G \longrightarrow \mathbb{C}^{\times}$.

b) The map $f \longmapsto \int_{G/H} dm(\dot{x}) \int_H f(xh)\, dh \qquad (f \in C_c(G))$

is a relatively invariant measure μ on G (we have denoted by dh a Haar measure on H). More precisely,

$${}_sf(x) = f(s^{-1}x) \implies \mu({}_sf) = \chi(s)\mu(f) \qquad (s \in G).$$

c) The measure $\mu/\chi : f \longmapsto \mu(f/\chi)$ is a Haar measure dx of G. Thus we can write $\mu = \chi\, dx$ and

$$\int_{G/H} dm(\dot{x}) \int_H f(xh)\, dh = \int_G f(x)\chi(x)\, dx \qquad (f \in C_c(G)).$$

d) Replacing f by one of its right translates f_t for some $t \in H$ ($f_t(x) = f(xt)$), conclude from the preceding equality that

$$\Delta_H = \chi \cdot \Delta_G \quad .$$

In particular, Δ_H has an extension $\chi\Delta_G : G \longrightarrow \mathbb{C}^{\times}$ which is a character.

e) Let $G = Gl_2(\mathbb{R})$, H = upper triangular subgroup in G. Show that there is <u>no</u> relatively invariant measure on G/H. (Observe that any character of $Gl_2(\mathbb{R})$ is trivial on the subgroup $Sl_2(\mathbb{R})$ and thus Δ_H cannot be extended as character of G. Remember that $Sl_2(\mathbb{R})$ is the commutator subgroup of G.)

f) Let G and H be as in e). Identifying the quotient G/H with the projective line via

$$\begin{pmatrix} a & b \\ c & d \end{pmatrix} \bmod H \longmapsto \text{line generated by } \begin{pmatrix} a \\ c \end{pmatrix} \text{ in } \mathbb{R}^2$$
$$\text{or } x = a/c \in \mathbb{R} \cup \{\infty\},$$

show that the Lebesgue measure dx of $\mathbb{R}$ gives a quasi-invariant measure on the homogeneous space G/H.

2. *Let G be a locally compact group acting continuously and transitively on a locally compact space X. Show that if G is countable at infinity, the mappings*

$$G \longrightarrow X \ , \quad s \longmapsto s\cdot x \qquad (x \in X)$$

are open mappings. In particular, for any $x \in X$, X can be identified to the homogeneous space $G/Stab(x)$ of G. (Observe that if K is any compact neighbourhood of e in G, G is a countable union of translates of K. Since X is a Baire space, $K\cdot x$ will contain an interior point.)

3. *Let G be a locally compact group and H, M two closed subgroups with a product HM open in G. We wish to determine the restriction m of a Haar measure dx of G on HM. We assume that $H \times M$ is countable at infinity and let this group act on G by*

$$(h,m)\cdot x = hxm^{-1} \qquad (h \in H,\ m \in M,\ x \in G).$$

Thus, the open set HM is the orbit of $e \in G$ under this action of $H \times M$. We identify HM with the homogeneous space

$$H \times M/A \ , \quad A = \{(a,a) : a \in H \cap M\} \subset H \times M$$

(Ex.2 above).

a) Show that m is a relatively invariant measure on this homogeneous space. More precisely,

$$\int_{HM} f(hxm^{-1})\, dx = \Delta_G(m) \int_{HM} f(x)\, dx \qquad (f \in C_c(HM)).$$

b) Prove that

$$\int_{HM} f(x)\, dx = \int_{H \times M} f(hm)\, \Delta_G(m)\, \Delta_H(m)^{-1}\, dh\, dm \quad .$$

4. *Let G be the locally compact group $\mathbb{R} \times \mathbb{R}_{discrete}$, m a Haar measure of G, and H the closed subgroup $\{0\} \times \mathbb{R}_{discrete}$. Show that*

$$\sup_{K \text{ compact},\ K \subset H} m(K) \neq \inf_{U \text{ open},\ U \supset H} m(U) \quad .$$

(In fact, Haar measures are regular Borel measures and equality would hold for all Borel parts H' for which both quantities are finite.)

13 CONTINUITY PROPERTIES OF REPRESENTATIONS

Let V be a locally convex (separated) topological vector space and End(V) be the space of continuous linear mappings $V \to V$. We also denote by Aut(V) = Gl(V) the group of invertible elements of End(V). Thus $A \in Gl(V)$ means that A is an invertible linear map $V \to V$ and both A and A^{-1} are continuous.

Any homomorphism π of a group G into Gl(V) defines a linear action (still denoted π) of G on V

$$\pi : G \to Gl(V) \ , \quad G \times V \to V \ , \quad (s,v) \mapsto \pi(s)\,v \ .$$

(13.1) Definition. Let G be a locally compact group and V as above. A representation π of G in V is a homomorphism $G \to Gl(V)$ for which the corresponding action $G \times V \to V$ is continuous.

It is useful to introduce the notion of *separately continuous* linear action when all

$$s \mapsto \pi(s)\,v \qquad (v \in V),$$

$$v \mapsto \pi(s)\,v \qquad (s \in G)$$

are continuous. The second condition implies $\pi(s) \in Gl(V)$ and thus, any separately continuous linear action defines a homomorphism $\pi : G \to Gl(V)$.

(13.2) Proposition. Let G be a locally compact group, V a barrelled space and π a separately continuous linear action of G in V. Then π is a representation.

Proof. The Banach-Steinhaus theorem shows that

i) for every compact subset K of G, the set $\pi(K) \subset End(V)$ is equicontinuous.

Since G is locally compact, this property is equivalent to

ii) there exists a compact neighbourhood Ω of $e \in G$ with $\pi(\Omega)$ equicontinuous in End(V).

Then take $s = s_o x$ near $s_o \in G$ and v near $v_o \in V$ and write

$$\pi(s)v - \pi(s_o)v_o = \pi(s_o)\left[\pi(x)(v - v_o) + \pi(x)v_o - v_o\right].$$

Thus continuity of the action follows from separate continuity and equicontinuity of the $\pi(x)$ $(x \in \Omega)$. q.e.d.

A unitary representation π of G in V is a representation for which V is a Hilbert space and each $\pi(s)$ $(s \in G)$ is a unitary operator. Thus, a unitary representation π of G in a Hilbert space H is a homomorphism $\pi : G \dashrightarrow U(H)$ into the unitary subgroup $U(H) \subset Gl(H)$ such that all maps $s \mapsto \pi(s)v$ $(v \in H)$ are continuous on G.

By the above proposition (13.2), any separately continuous linear action of G in a Hilbert space is a representation. But when all operators $\pi(s)$ $(s \in G)$ are unitary, $\|\pi(s)\| = 1$ shows that $\pi(G)$ is equicontinuous (in fact, U(H) is an equicontinuous set of operators) and the condition ii) of the above proof is trivially satisfied. Thus, the global continuity of the action defined by π follows without using the Banach-Steinhaus theorem.

(13.3) <u>Proposition</u>. <u>Let</u> G <u>be a locally compact group</u>, H <u>a Hilbert space</u> <u>and</u> U(H) <u>the unitary subgroup of</u> Gl(H). <u>Then any homomorphism</u>

$$\pi : G \to U(H)$$

<u>for which all</u> $s \mapsto (v \mid \pi(s)v)$ $(v \in H)$ <u>are continuous at the neutral element</u> $e \in G$, <u>is a unitary representation of</u> G.

<u>Proof</u>. Observe that for s and $t \in G$

$$\|\pi(s)v - \pi(t)v\|^2 = \|\pi(t)^{-1}\pi(s)v - v\|^2 =$$

$$= (\pi(x)v - v \mid \pi(x)v - v) = 2(v \mid v) - 2\,\mathrm{Re}\,(v \mid \pi(x)v)$$

where $x = t^{-1}s$. Thus the proposition follows. q.e.d.

REMARKS

1) Since weak and strong topologies coïncide on the unitary group of a Hilbert space (cf. Dieudonné 1969, Chap.12, sec.15 p.79), continuity of a map $s \mapsto \pi(s)v$ certainly follows from continuity of all $s \mapsto (w \mid \pi(s)v)$ $(w \in H)$. Moreover, on any bounded subset of H, the weak topology can be defined by any total subset of H. Thus it would be enough to check the continuity of all $s \mapsto (e_i \mid \pi(s)v)$ where (e_i) denotes an orthonormal basis of H. However, one should observe

that U(H) is not closed for the strong topology on the unit ball of End(H) (this point seems to have escaped Dieudonné's attention, loc. cit.). If $(A_n) \subset U(H)$ is a strongly convergent sequence of operators, its limit $A = \lim A_n$ (defined by $A(v) = \lim A_n(v)$) is an isometry but is not always surjective (take for example for A_n the operator defined by

$$A_n(e_i) = e_{i+1} \text{ for } 0 \leq i < n , \quad A_n(e_n) = e_o$$

$$\text{and } A_n(e_i) = e_i \quad \text{for } i > n$$

in $H = \ell^2(\mathbb{N})$: the strong limit of this sequence is the shift operator).

2) It is possible to give weaker conditions for a homomorphism $\pi : G \to Gl(H)$ to be a representation . For example, if H is a separable Hilbert space, it is enough to check that the mappings

$$s \mapsto (w \mid \pi(s)v) \qquad (w, v \in H)$$

are measurable. In this case, the mappings $s \mapsto \pi(s)v$ $(v \in H)$ are called weakly measurable, and weak measurability implies strong measurability (Bourbaki 1959, Prop.12 p.21). For these properties, the reader should consult Gaal 1973 (Th.1 p.304, Prop.2 p.305 and Th.3 p.306). But here is an example where weak measurability is not sufficient to imply measurability (and strong continuity). We take $G = \mathbb{R}$ and H a Hilbert space with orthonormal basis $(e_t)_{t \in \mathbb{R}}$. The homomorphism

$$\pi : \quad G = \mathbb{R} \to U(H) , \quad s \mapsto \pi(s) ,$$

is defined by $\pi(s)(e_t) = e_{t+s}$. The mappings

$$G \to H , \quad s \mapsto \pi(s)e_t = e_{t+s}$$

are not measurable (although they are weakly measurable) and π is not a unitary representation in our sense (H is *not* separable so that the general results quoted above do not apply).

3) For a discussion of the continuity properties of representations, a general reference is Bourbaki 1963 (sec.2 p.128). A somewhat easier treatment is given in Borel 1972 (p.16).

Irreducibility of representations is defined as in sec.2 (p.14) and Prop. (2.1) is still valid.

EXAMPLES

1) Let G be a commutative locally compact group. Schur's lemma shows that the unitary irreducible representations of G have dimension 1 and can thus be identified with the continuous homomorphisms

$$\chi \ : \ G \longrightarrow U_1(\mathbb{C}) \subset \mathbb{C}^{\times} \ .$$

These are the *characters* of G. The set $\hat{G}$ of characters of G is a group (pointwise multiplication) and in fact a locally compact group for the topology of uniform convergence on compact sets. This *dual* $\hat{G}$ of G again has a dual and the Pontryagin duality shows that

$$(\hat{G})^{\wedge} \quad \text{is canonically isomorphic to } G \ .$$

For example if $G = \mathbb{R}$, the characters $\mathbb{R} \longrightarrow U_1(\mathbb{C})$ can all be written in the form

$$t \longmapsto e^{ist} \qquad\qquad (s \in \mathbb{R})$$

and the dual $\hat{G} = \hat{\mathbb{R}}$ is isomorphic to $\mathbb{R}$. The regular representation of $\mathbb{R}$ in $L^2(\mathbb{R})$ is a unitary representation (cf. infra., since this group is commutative, its left and right regular representations coîncide). and it is easy to construct non-trivial closed invariant subspaces of this representation. For this purpose, let us denote by

$$F : \ L^2(\mathbb{R}) \ \longrightarrow \ L^2(\mathbb{R})$$

the Fourier transform (it is a unitary operator defined on $L^1(\mathbb{R}) \cap L^2(\mathbb{R})$ by the integral formula

$$(Ff)(s) \ = \int_{\mathbb{R}} f(t)\, e^{-ist}\, dt \quad) \ .$$

Then for any (measurable) subset $A \subset \mathbb{R}$, put

$$H_A \ = \left\{ f \in L^2(\mathbb{R}) \ : \ \ Ff \text{ vanishes outside } A \right\} .$$

Thus $H_A = \{0\}$ when A is negligible and $H_A = L^2(\mathbb{R})$ when $\mathbb{R} - A$ is negligible. If f_a denotes a translate of f, the formula

$$F(f_a)(s) \ = \ e^{-ias}\, F(f)(s)$$

shows that all subspaces H_A are invariant under translations. But it can be shown that $L^2(\mathbb{R})$ contains no invariant irreducible subspace.

2) Let G be a locally compact group acting continuously on a locally compact space X admitting a quasi-invariant positive measure m, say

$$dm(s^{-1}x) = |c_s(x)|^2 \, dm(x) \qquad (s \in G, \ x \in X).$$

To make integration theory easier to handle, let us assume that X is countable at infinity (there is a countable fundamental system of compact sets in X). We also assume that the functions c_s are continuous on X and tend to 1 uniformly on compact sets when $s \to e \in G$. Then the formulas

$$(\pi(s)f)(x) = c_s(x) \, f(s^{-1}x)$$

define a unitary representation π of G in $H = L^2(X,m)$. First of all,

$$\int_X |f(s^{-1}x)|^2 |c_s(x)|^2 \, dm(x) = \int |f(s^{-1}x)|^2 \, dm(s^{-1}x) =$$

$$= \int |f(x)|^2 \, dm(x) = \|f\|^2$$

proves that $\pi(s)$ is a unitary operator in H for all $s \in G$. Then, the separate continuity of the action associated to π can be verified as in sec.2 p.17, starting with the continuity of the maps

$$G \to H, \quad s \mapsto \pi(s)f \quad \text{with} \quad f \in C_c(G).$$

Examples of this kind often arise: homogeneous spaces G/H always admit quasi-invariant measures (with respect to left translations). We shall have to consider the case of $G = Sl_2(\mathbb{R})$, X = upper half-plane consisting of the $z = x + iy$ with $y = \mathrm{Im}(z) > 0$ and positive measures

$$m = m_k = y^{k-2} \, dx \, dy \qquad (k \in \mathbb{N}).$$

It is of course possible to generalize this kind of construction (cf. Bourbaki 1963 Ex.13, p.199 for example).

14 REPRESENTATIONS OF G AND OF $L^1(G)$

Let G be a locally compact group and $L^1(G)$ the Banach space of integrable functions on G with respect to a fixed Haar measure on G. Thus $L^1(G)$ is the completion of the vector space $C_c(G)$ of (complex) continuous functions $f : G \longrightarrow \mathbb{C}$ vanishing outside some compact set of G, with the norm

$$\|f\| = \|f\|_1 = \int_G |f(s)|\, ds \quad .$$

The formula

$$(14.1) \qquad f * g\,(t) = \int_G f(s)\, g(s^{-1}t)\, ds$$

defines the convolution product first on $C_c(G)$ and then on $L^1(G)$ by continuity :

$$\int_G |f * g(t)|\, dt \leqslant \iint_{G \times G} |f(s)\, g(s^{-1}t)|\, ds\, dt =$$

$$= \int ds\, |f(s)| \int dt\, |g(s^{-1}t)| = \|g\| \int ds\, |f(s)| = \|f\|\, \|g\|$$

(first for f and g in $C_c(G)$ and then also for f and g in $L^1(G)$ by continuity) shows that the space $L^1(G)$ is a Banach algebra with respect to convolution

$$\|f * g\| \leqslant \|f\|\, \|g\| \qquad (f,\, g \in L^1(G)\,).$$

There is also an *involution* $f \mapsto f^*$ on $L^1(G)$. It is defined by

$$(14.2) \qquad f^*(x) = \overline{f(x^{-1})}\, \Delta_G(x)^{-1} \quad .$$

Indeed, (12.4) shows immediately that

$$(14.3) \qquad \begin{cases} (f^*)^* = f \quad , \\ \|f^*\| = \|f\| \end{cases} \qquad (f \in L^1(G)\,).$$

Also observe that

$$f^* * g\,(e) = \int f^*(s)\, g(s^{-1})\, ds = \int \overline{f(s^{-1})}\, g(s^{-1})\, d(s^{-1}) =$$

$$= (f \mid g) \qquad \text{for } f \text{ and } g \text{ in } L^1(G) \cap L^2(G)\,.$$

When G is not discrete, $L^1(G)$ has no unit for convolution. But here is a substitute.

(14.4) Lemma. The convolution algebra $L^1(G)$ always has an approximate unit. More precisely, there is a family $(u_i) \subset L^1(G)$ with

$$\|u_i\| \leqslant 1 \ , \ \|u_i * f - f\| \to 0 \ , \ \|f * u_i - f\| \to 0$$

for each $f \in L^1(G)$. In fact, we can even take $0 \leqslant u_i \in L^1(G)$ and $\|u_i\| = 1$.

Proof. Let indeed (V_i) be a fundamental system of neighbourhoods of $e \in G$ (parametrized by some ordered set I for which $j \geqslant i \iff V_j \subset V_i$) and take $0 \leqslant u_i \in C_c(G)$ with

$$\|u_i\| = \int u_i(s)\, ds = 1 \quad \text{and} \quad u_i \text{ vanishes outside } V_i \ .$$

Then the lemma follows. In fact, if $f \in C_c(G)$, $u_i * f \to f$ and $f * u_i \to f$ uniformly on G. q.e.d.

Let us observe that when the group G is unimodular, the definition (14.1) of the convolution product in $L^1(G)$ can also be given equivalently by

$$(14.5) \qquad f * g(t) = \int_G f(ts^{-1})\, g(s)\, ds \qquad (G \text{ unimodular}) \ .$$

In this case, we also have

$$f^*(x) = \overline{f(x^{-1})} \quad (\text{i.e. } f^* = \check{\bar{f}}) \ .$$

The presence of the modular function in the definition of f^* in general is best explained by the consideration of this involution on the Banach space (algebra) $M^1(G)$ consisting of bounded measures on G. Using a fixed Haar measure m of G, there is an embedding

$$L^1(G) \hookrightarrow M^1(G) \ , \quad f \mapsto f\cdot m \ ,$$

which is an isometry. If $n \in M^1(G)$ is a bounded measure on G, one defines a bounded measure $|n|$ by

$$|n|(f) = \operatorname*{Sup}_{|g| \leqslant f} |n(g)| \qquad \text{for } 0 \leqslant f \in C_c(G) \ .$$

One can then check

$$|f\cdot n| \leqslant |f|\cdot|n| \qquad (f \in L^1(G),\ n \in M^1(G))$$

$$|n * n'| \leqslant |n| * |n'| \qquad (n, n' \in M^1(G)).$$

The Fubini theorem shows that

$$m(A) = 0 \implies |f\cdot m * n|(A) \leqslant |f\cdot m| * |n|(A) = 0$$

for $f \in L^1(G)$ and $n \in M^1(G)$. Then, the Radon-Nikodym theorem asserts that $f\cdot m * n$ is a multiple of the measure m

$$f\cdot m * n = g\cdot m \quad \text{for some} \quad g \in L^1(G) .$$

Thus, $L^1(G)$ can be identified with a two-sided ideal of $M^1(G)$ and the involution $n \mapsto \bar{n}^\vee$ gives, on the image of $L^1(G)$

$$(f\cdot m)^* = \overline{(f\cdot m)}^\vee = \bar{f}^\vee m^\vee = \bar{f}^\vee \Delta_G^{-1}\cdot m \qquad (f \in L^1(G)$$

by (12.4). One of the advantages of working with $M^1(G)$ is the possibility of embedding G in this algebra : for $s \in G$, we denote by ε_s the evaluation measure $f \mapsto \varepsilon_s(f) = f(s)$ so that

$$\varepsilon_s * \varepsilon_t = \varepsilon_{st}$$

and

$$G \to M^1(G) , \quad s \mapsto \varepsilon_s$$

is an injective homomorphism. However, to remain more self-contained, we shall only make a limited use of this bigger convolution algebra .

For any representation π of G in a (quasi-complete) space V, we can define the operators

$$v \mapsto \tilde{\pi}(f)v = \int_G f(s)\ \pi(s)v\, ds \qquad (f \in C_c(G)).$$

When π is unitary, one can also define the operators

$$\pi^1(f) = \int_G f(s)\ \pi(s)\, ds$$

by the same formula for all $f \in L^1(G)$ (more generally, for $n \in M^1(G)$, one can define $\pi^1(n) = \int \pi(s)\, dn(s)$). One checks without difficulty

$$(14.6) \qquad \begin{cases} \pi^1(f * g) = \pi^1(f)\, \pi^1(g) , \\ \pi^1(f^*) = \pi^1(f)^* , \ \|\pi^1(f)\| \leqslant \|f\| , \\ \pi^1(\varepsilon_s * f) = \pi^1({}_s f) = \pi(s)\, \pi^1(f) \end{cases}$$

for f and $g \in L^1(G)$ and $s \in G$.

Thus we obtain a representation $\pi^1 : L^1(G) \to \mathrm{End}(V)$.

Moreover with an approximate unit (u_i) as defined in the proof of (14.4), for each $\varepsilon > 0$, we can find an index i such that

$$u_i(x) \neq 0 \implies \|\pi(x)v - v\| \leq \varepsilon \quad \text{(for a fixed } v \in V)$$

and thus

$$\|\pi^1(u_i)v - v\| \leq \int u_i(x)\|\pi(x)v - v\|\, dx \leq \varepsilon \quad .$$

This proves that for each $v \in V$, $\pi^1(u_i)v \longrightarrow v$. In particular, the vectors $\pi^1(f)v$ $(f \in L^1(G)$, $v \in V)$ generate V topologically.

Quite generally, if A is a $*$-algebra ($*$ here refers to an involution, not to convolution...) , H a Hilbert space and ρ a $*$-homomorphism $A \to \mathrm{End}(H)$, we have an orthogonal direct sum

$$H = H_o \oplus H_1$$

with

H_o = intersection of the kernels of the $\rho(a)$ $(a \in A)$,

H_1 = closure of the space generated by the $\rho(a)v$ $(a \in A, v \in V)$.

This follows immediately from the formulas

$$(w \mid \rho(a)v) = (\rho(a)^* w \mid v) = (\rho(a^*)w \mid v)$$

for $a \in A$ and $v, w \in H$. We say that ρ is <u>non-degenerate</u> when $H_o = \{0\}$. Thus, for any unitary representation of G in a Hilbert space H, the $*$-representation π^1 of $A = L^1(G)$ is non-degenerate. There is a converse to this result.

(14.7) <u>Theorem</u>. <u>Let</u> G <u>be a locally compact group</u>, H <u>a Hilbert space and</u> ρ <u>a non-degenerate</u> $*$-<u>representation</u> $L^1(G) \to \mathrm{End}(H)$. <u>Then there is a unique unitary representation</u> $\pi : G \to \mathrm{Gl}(H)$ <u>such that</u> $\rho = \pi^1$.

The uniqueness of π follows from the third formula of (14.6)

$$\rho = \pi^1 \text{ and } \pi^1({}_sf) = \pi(s)\,\pi^1(f) \implies \rho({}_sf)v = \pi(s)\,\rho(f)v$$

since, by assumption, the vectors $\pi^1(f)v = \rho(f)v$ ($f \in L^1(G)$, $v \in H$) generate H . At the same time, this formula shows that the only possible definition of $\pi(s)$ is an extension of

$$\rho(f)v \longmapsto \rho({}_sf)v \quad .$$

We have to check that this association is well defined and has all required properties. This is based on the following general result.

(14.8) Lemma. Any $*$-homomorphism $L^1(G) \xrightarrow{\rho} \mathrm{End}(H)$ is contracting and hence, continuous.

Proof of this lemma. It is known that if $T \in \mathrm{End}(H)$ is a hermitian operator, $\|T\|$ is given by the spectral radius of T. By definition, this spectral radius is

$$\text{spec rad}\, T = \lim_{k\to\infty} \|T^k\|^{1/k} .$$

But End(H) is a stellar algebra (C*-algebra)

$$\|T^2\| = \|T^*T\| = \|T\|^2$$

so that by induction on n, we get $\|T^{2^n}\| = \|T\|^{2^n}$ and the above limit is easily computed with $k = 2^n$:

$$\text{spec rad}\, T = \|T\| \qquad \text{(T hermitian)}.$$

Now it $f = f^* \in L^1(G)$, $T = \rho(f)$ is a hermitian operator and thus

$$\|\rho(f)\| = \text{spec rad}\, \rho(f) \leqslant \text{spec rad}\, f \leqslant \|f\|$$

($f - \lambda 1$ invertible for convolution in $L^1(G)$ implies $\rho(f) - \lambda 1$ invertible in End(H), and hence $\text{spec}\, \rho(f) \subset \text{spec}\, f$). Now when $g \in L^1(G)$ is arbitrary, $f = g^* * g$ satisfies $f^* = f$ as above and

$$\|\rho(g)\|^2 = \|\rho(g)^*\rho(g)\| = \|\rho(f)\| < \|f\| \leqslant \|g^*\|\,\|g\| = \|g\|^2$$

whence the conclusion. (As the proof has shown, $L^1(G)$ could be replaced by any $*$-algebra A and End(H) by any stellar algebra B in this lemma : cf. Dixmier 1964, (1.3.7) p.7 .)

End of proof of (14.7). By lemma (14.4), for $f \in L^1(G)$,

$$u_i * f \longrightarrow f \text{ in } L^1(G) ,$$

and the same convergence will hold for left translates

$$_s u_i * f = {}_s(u_i * f) \longrightarrow {}_s f \quad \text{in } L^1(G) .$$

Now lemma (14.8) shows that $\rho({}_s u_i)\rho(f) \rightarrow \rho({}_s f)$ and hence, we can define

$$\pi(s)\, w = \lim_i \rho({}_s u_i)\, w$$

for all linear combinations w of special vectors $\rho(f)v$ ($f \in L^1(G)$, $v \in V$).

Still by (14.8), $\|\rho({}_s u_i)\| \leqslant 1$, hence $\|\pi(s)\| \leqslant 1$ and $\pi(s)$ has a unique continuous (contracting) extension to the whole space H. Obviously

$$\pi(e) = \text{id.}, \quad \pi(st) = \pi(s)\,\pi(t)$$

and thus $\pi(s^{-1}) = \pi(s)^{-1}$. Now

$$\|\pi(s)\| \leqslant 1 \text{ and } \|\pi(s)^{-1}\| = \|\pi(s^{-1})\| \leqslant 1$$

implies that $\pi(s)$ is unitary ($s \in G$). The homomorphism $\pi : G \longrightarrow U(H)$ is a representation since, by definition

$$s \longmapsto \pi(s)\,w = \pi(s)\rho(f)\,v = \rho({}_s f)\,v$$

is continuous for a total system of vectors w (take the vectors w of the form $\rho(f)\,v$ for some $v \in H$ and $f \in C_c(G)$, in order to ensure ${}_s f \longrightarrow {}_t f$ in $L^1(G)$, when $s \to t$ in G as on p. 17). We have to check that with our definition of π, $\pi^1 = \rho$. But

$$f * g = \int f(s)\,(\varepsilon_s * g)\,ds$$

leads to

$$\rho(f)\,\rho(g) = \rho(f * g) = \int f(s)\,\rho(\varepsilon_s * g)\,ds =$$

$$= \int f(s)\,\pi(s)\,ds\;\rho(g) = \pi^1(f)\,\rho(g) \quad .$$

Since ρ is non-degenerate, this implies $\rho(f) = \pi^1(f)$. q.e.d.

If π is a unitary representation of G in H, the preceding construction (14.7) can be made in any closed invariant subspace of H. Thus π and π^1 have the same closed invariant subspaces. In particular

(14.9) Corollary. If π is a unitary representation of G in H,

$$\pi \text{ irreducible} \iff \pi^1 \text{ irreducible}.$$

From now on, we can safely write π instead of π^1.

As for compact groups, one could still extend π to the stellar algebra of G. For this purpose, one should define

$$\|f\|_* = \sup_{\pi \text{ unitary}} \|\pi(f)\| \qquad (f \in L^1(G))$$

and define $C^*(G)$ as completion of $L^1(G)$ for this norm. By definition, the embedding $L^1(G) \longrightarrow C^*(G)$ has a dense image and any unitary representation π of G (in some Hilbert space H) has an extension

$$\pi^* \quad : \quad C^*(G) \longrightarrow \mathrm{End}(H) \quad .$$

One can prove that

$$\pi \text{ irreducible} \quad \Longrightarrow \quad \pi^* \textit{ algebraically} \text{ irreducible}$$

(cf. Dixmier 1964, (2.8.4) p.45).

In particular, the left regular representation of G can be so extended

$$\ell^* \quad : \quad C^*(G) \longrightarrow \mathrm{End}(H) \quad \text{with} \quad H = L^2(G) \; .$$

The image of this homomorphism is a stellar algebra $C^*_{red}(G)$ which is, in general, a proper quotient of $C^*(G)$. We shall not pursue these topics any further.

As we have already seen with the regular representation of the group $G = \mathbb{R}$, a unitary representation of a locally compact group need not have invariant irreducible subspaces. Here is however an interesting case.

(14.10) <u>Theorem</u>. <u>Let</u> π <u>be a unitary representation of a locally compact group</u> G <u>in a Hilbert space</u> H <u>such that the operators</u> $\pi(u_i)$ <u>are compact for all elements of an approximate unit</u> $(u_i) \subset C_c(G)$. <u>Then</u> H <u>is a Hilbert sum of irreducible subspaces with only finitely many of them being equivalent to a given one</u>.

<u>Proof</u>. For $f \in C_c(G)$, the functions $f * u_i$ converge uniformly (hence in $L^1(G)$ by compactness of all supports) and thus $\pi(f * u_i) = \pi(f)\,\pi(u_i)$ converges to $\pi(f)$ in the operator norm. Since any norm limit of compact operators is compact

$$\pi(f) \text{ is a compact operator for every } f \in C_c(G) \; .$$

Since $\pi(u_i)v \longrightarrow v$ $(v \in H)$, there is certainly an index j with $\pi(u_j) \neq 0$. Taking

$$f = \tfrac{1}{2}(u_j^* + u_j) \quad \text{or} \quad f = \frac{1}{2i}(u_j - u_j^*) \quad ,$$

we see that there is a function $f = f^* \in C_c(G)$ with $0 \neq \pi(f)$ hermitian. Since this operator is compact, hermitian and $\neq 0$, it must have a non-zero eigenvalue λ. The corresponding eigenspace

$$H_\lambda \; = \; \mathrm{Ker}(\pi(f) - \lambda 1_H) \quad \neq \{0\}$$

is finite-dimensional. For $a \in H$, we denote by $H(a)$ the closed subspace generated by the $\pi(f)a$ $(f \in L^1(G))$. This subspace $H(a)$ is always invariant under π. We need a lemma.

(14.11) Lemma. If H_1 is a closed invariant subspace of $H(a)$, then the orthogonal projection a_1 of a in H_1 belongs to the eigenspace H_λ and $H_1 = H(a_1)$. Moreover, for two closed invariant subspaces $H_j \subset H(a)$

$$H_1 \subsetneqq H_2 \iff H_1 \cap H_\lambda \subsetneqq H_2 \cap H_\lambda .$$

To prove this lemma, we notice that the orthogonal projector P_1 on H_1 commutes to all operators $\pi(f)$ $(f \in L^1(G))$. Taking for f the preceding hermitian element of $L^1(G)$, we see that

$$\pi(f)a_1 = \pi(f)P_1 a = P_1\pi(f)a = P_1\lambda a = \lambda a_1 .$$

Now, the vectors $\pi(f')a$ $(f' \in L^1(G))$ are total in H (if $a \neq 0$) so that their projections

$$P_1\pi(f')a = \pi(f')a_1 \qquad (f \in L^1(G))$$

also make up a total system in $H_1 = H(a_1)$. Finally, if $H_1 \cap H_\lambda = H_2 \cap H_\lambda$, the orthogonal projection a_1 of a_2 on H_1 is equal to a_2 : $a_2 = a_1$ and thus $H_2 = H_1$ by what we have already proved. Thus the lemma follows.

We can now finish the proof of (14.10). From the lemma (14.11) we infer that the set of invariant subspaces of $H(a)$ has an artinian property: any decreasing sequence of closed invariant subspaces of $H(a)$ must be stationary (because the same property holds in the finite-dimensional space $H(a) \cap H_\lambda$). In particular, $H(a)$ contains an irreducible subspace (if $a \neq 0$). The preceding argument can of course be repeated in any closed invariant subspace of H (the hypotheses of (14.10) will still be satisfied for the restriction of π to such a subspace):
any non-zero closed invariant subspace of H contains an irreducible subspace. Zorn's lemma applies to the set of families of orthogonal closed irreducible subspaces of H and there exists a maximal such family $(H_i)_I$. The orthogonal of the Hilbert sum $\widehat{\bigoplus_I} H_i$ is closed and invariant. If this subspace were non-zero, we could find an irreducible subspace in it and $(H_i)_I$ would not be maximal. Thus $H = \widehat{\bigoplus_I} H_i$ is a Hilbert

sum of irreducible subspaces. Select one irreducible subspace H_1 and collect all equivalent subspaces H_j $(j \in J)$. There exists a hermitian function $f \in C_c(G)$ with $\pi(f)$ non-zero in H_1. Thus, we could take a non-zero eigenvalue of $\pi(f)$ in H_1. Since the operator $\pi(f)$ induces the multiplication by λ in the corresponding subspace $\bigoplus_J H_{j,\lambda}$ and is compact, this last space must be finite-dimensional : the index set J must be finite. q.e.d.

The preceding proof shows that any representation of G in a Banach space such that all operators $\pi(f)$ $(f \in C_c(G))$ are compact has an irreducible subspace (if $\neq \{0\}$) and hence a generalized composition series with irreducible quotients.

Perhaps one should observe that the conditions of the theorem (14.10) are satisfied if G is a compact group and $\pi = \ell$ is the left regular representation of G in $L^2(G)$ (by the exercise 2 of sec.6, $\ell^1(f) = f *$ is an integral operator to which lemma 1 of sec.4 applies). More generally, (14.10) admits the following interesting application.

(14.12) <u>Application</u>. <u>Let</u> Γ <u>be a discrete subgroup of a locally compact unimodular group</u> G. <u>Then the left regular representation of</u> G <u>in</u> $L^2(G/\Gamma)$ <u>decomposes as a Hilbert sum of irreducible ones</u>, <u>each occuring with finite multiplicities</u>.

<u>Proof</u>. We have to check that the assumptions of (14.10) are satisfied. Since Γ is discrete, it is unimodular and G/Γ carries an invariant measure (12. 8) from which the space $L^2(G/\Gamma)$ is built. For $f \in C_c(G)$ and $\varphi \in C_c(G/\Gamma)$, we have

$$\ell(f)\varphi(\dot{x}) = \int_G f(s)\,\varphi(s^{-1}\dot{x})\,ds = \int f(s^{-1})\,\varphi(s\dot{x})\,ds =$$

$$= \int f(xs^{-1})\,\varphi(\dot{s})\,ds =$$

$$= \int_{G/\Gamma} d\dot{s}\;\varphi(\dot{s}) \sum_{\gamma\in\Gamma} f(x\gamma s^{-1}) \quad .$$

Hence $\ell(f)$ is the integral operator on G/Γ with kernel

$$k_f(\dot{x},\dot{y}) = \sum_{\gamma\in\Gamma} f(x\gamma y^{-1}) \quad \text{on} \quad G/\Gamma \times G/\Gamma \quad .$$

The preceding sum converges : let K = Supp(f) (this is a compact set)

$$f(x\gamma y^{-1}) \neq 0 \implies x\gamma y^{-1} \in K \cap x\Gamma y^{-1}$$

But the set $K \cap x\Gamma y^{-1}$ is compact and discrete, hence finite, so that for each given couple (x,y), the sum defining the kernel k_f is in fact a finite sum. More generally, if Ω is a compact neighbourhood of $e \in G$,

$$\Omega K \Omega^{-1} \cap x\Gamma y^{-1} \quad \text{is finite}$$

and the sum defining k_f has only finitely many non-zero terms on $\Omega x \times \Omega y$. It is thus uniformly convergent on all compact sets and k_f is a continuous kernel. Thus (14.12) follows from the following general result and (8.2).

(14.13) Theorem. Let X be a locally compact space, m a positive measure on X and $k \in L^2(X \times X, m \otimes m)$. Then the integral operator

$$f \mapsto Kf \ , \quad (Kf)(x) = \int_X k(x,y)\, f(y)\, dm(y)$$

is a Hilbert-Schmidt operator in $L^2(X,m)$ and

$$\|K\|_2^2 = \iint_{X \times X} |k(x,y)|^2\, dm(x)\, dm(y) \quad .$$

Proof. Fubini's theorem shows that the functions $k(x,.) \in L^2(X,m)$ except for a negligible set of $x \in X$. Thus, for nearly all $x \in X$

$$k(x,.)f \in L^1(X,m) = L^1 \qquad (f \in L^2) \ ,$$

and $(Kf)(x) = (\bar{f} \mid k(x,.))$ is well defined for the same values of x. The Cauchy-Schwarz inequality gives

$$|(Kf)(x)| \leqslant \|k(x,.)\|_2 \, \|f\|_2 \ .$$

By integration of the square of that quantity, we get

$$\|Kf\|_2^2 \leqslant \|f\|_2^2 \int dm(x) \int dm(y)\, |k(x,y)|^2 \ .$$

This shows that

$$Kf \in L^2 \ , \quad f \mapsto Kf \text{ continuous } L^2 \to L^2$$

and

$$\|K\|^2 \leqslant \int |k(x,y)|^2\, dm(x)\, dm(y) \qquad \text{(uniform norm)},$$

but we need much more... Take an orthonormal basis (e_i) in $H = L^2$ and denote by $(\bar{e}_i)$ the orthonormal basis with $\bar{e}_i(x) = \overline{e_i(x)}$. For nearly all $x \in X$, we have an expansion (in L^2)

$$k(x,.) = \sum a_i(x)\, \bar{e}_i$$

with coefficients given by the usual scalar products

$$a_i(x) = (\bar{e}_i \mid k(x,.)) = Ke_i\,(x) \quad .$$

The Parseval identity is

$$\| k(x,.)\|_2^2 = \sum |Ke_i\,(x)|^2 \quad .$$

Integrating this identity (valid outside a negligible set of $x \in X$), we get

$$\infty > \iint dm(x)\, dm(y)\, |k(x,y)|^2 = \sum \| Ke_i\|_2^2 = \| K\|_2^2 \ .$$

q.e.d.

EXERCISES

1. Show that $L^1(G)$ is not a stellar algebra in general.
Hint: Show that for $f \in L^1(G)$,

$$\| f^* * f \| = \left| \int f(x)\, dx \right|^2 \quad (\leqslant \|f\|^2)$$

so that $f^* * f = 0$ as soon as the "average" of f is 0. But

$$\| f^* * f \| = \|f\|^2 \qquad \text{if} \quad f \geqslant 0 .$$

2. Let K and H be two integral operators in $L^2 = L^2(X,m)$ given by two kernels k and h on $X \times X$ (as in (14.13)) respectively.

a) Prove that the scalar product of these Hilbert-Schmidt operators is given by

$$\int p(x,x)\, dm(x)$$

where

$$p(x,y) = \int \overline{k(z,x)}\, h(z,y)\, dm(z)$$

is the kernel of K^*H.

b) Observe that $x \longmapsto p(x,x) = (k(.,x) \mid h(.,x))$ is in L^1 since it is the product of the two vector valued L^2-functions

$$x \longmapsto k(.,x) \quad ,$$

$$x \longmapsto h(.,x)$$

(use Bourbaki 1965, Int. Chap.IV, sec.6 Cor.1 de Th.2 p.208).

Remark. This exercise shows that the trace of the (trace class) operator K^*H is given by the integral over the diagonal of its kernel.

15 SCHUR'S LEMMA : UNBOUNDED VERSION

We have to review a few facts on unbounded operators in a Hilbert space. Here are three references for this theory

Riesz-Nagy 1975 (Chap. VIII),
Dieudonné 1969 (vol.2, Chap. XV, sec.11-13),
Lang 1975 (Appendix 2).

An unbounded operator T in a Hilbert space H is a linear map $D_T \longrightarrow H$ defined on a dense subspace $D_T \subset H$. To be more precise, one should speak of the unbounded operator (T, D_T). In particular, we consider as distinct, two unbounded operators having distinct domains of definition. An extension S of an unbounded operator T is an unbounded operator (S, D_S) with $D_S \supset D_T$ and $S|_{D_T} = T$. In this case, we simply write $S \supset T$.

(15.1) Definitions. a) An unbounded operator T is said to be closed when its graph ($\subset D_T \times H$) is closed in $H \times H$.

b) The operator T can be closed when it has a closed extension.

This second condition is satisfied when the closure of the graph of T is still a functional graph, namely when

$$x_n \to 0 \text{ and } T(x_n) \to y \quad \Longrightarrow \quad y = 0 .$$

When T can be closed, the smallest closed extension of T is the operator $\bar{T}$ having for graph the closure of the graph of T :

(15.2) $$x \in D_{\bar{T}} \iff \exists\,(x_n) \subset D_T \text{ with } x_n \to x \text{ and } \{Tx_n\} \text{ bounded.}$$

To see this, we first have to introduce the adjoint T^* of T. First, we define

$$D_{T^*} = \{x \in H : y \longmapsto (x \mid Ty) \text{ continuous on } D_T\} .$$

Then Riesz' theorem allows us to put $T^*x = z$ if

$$x \in D_{T^*} \text{ and } (x \mid Ty) = (z \mid y) = (T^*x \mid y) \qquad (y \in D_T).$$

Then we have

(15.3) $$D_{T^*} \text{ dense} \iff T \text{ can be closed} \implies \bar{T} = T^{**}$$

Coming back to (15.2), if $\{Tx_n\}$ is bounded, we can assume -extracting a subsequence if necessary- that Tx_n converges weakly to an element z. In particular, for all $y \in D_{T^*}$,

$$(Tx_n \mid y) \longrightarrow (z \mid y) , \quad (x_n \mid T^*y) \longrightarrow (x \mid T^*y)$$

whence $x \in D_{T^{**}}$ and $T^{**}x = z$. Thus (15.2) follows from (15.3).

For any operator T, T* is closed.

(15.4) Definitions. Let T be an unbounded operator.

a) T symmetric $\Longleftrightarrow$ $T \subset T^*$ (hence D_{T^*} dense and T can be closed)

b) T self-adjoint $\Longleftrightarrow$ $T = T^*$ (and $D_T = D_{T^*}$),

c) T essentially self-adjoint $\Longleftrightarrow$ T can be closed and $\bar{T} = \bar{T}^*$.

Thus symmetric operators are characterized by

$(y \mid Tx) = (Ty \mid x)$ for all x and y in D_T,

and essentially self-adjoint operators have a closure which is self-adjoint.
If T is an unbounded operator, T*T is canonically defined on

$$D_{T^*T} = \{x \in D_T : Tx \in D_{T^*}\} = D_T \cap T^{-1}D_{T^*} .$$

In fact, one can show that $(1 + T^*T)^{-1}$ is a continuous (bounded) operator: its norm is $\leqslant 1$ and it is positive (spectrum in $[0,1]$).

Here is a classical result

(15.5) T closed $\Longrightarrow$ T*T self-adjoint.

A normal operator T is a closed operator for which $T^*T = TT^*$ (this equality requires $D_{T^*T} = D_{TT^*}$). For example

$$T \text{ self-adjoint} \Longrightarrow T = T^{**} = \bar{T} \text{ closed} \Longrightarrow$$

$$\Longrightarrow T^2 = T^*T \text{ self-adjoint} \Longrightarrow T \text{ normal} .$$

There is a complete spectral theory for normal operators (in particular, for self-adjoint operators). In this theory, it is possible to define functions of T : in particular, $(T^*T)^{\frac{1}{2}}$ is the absolute value of T and

(15.6) Every closed operator T has a polar decomposition $T = UP$ where U is partially isometric ($U^*U = 1$) and $P = (T^*T)^{\frac{1}{2}}$ is self-adjoint and positive.

Here, T is an unbounded operator, but U is necessarily continuous with $\|U\| \leqslant 1$. When T is symmetric, $T \subset T^* \Longrightarrow \bar{T} \subset T^*$

and thus $\bar{T}$ is still symmetric (with $\bar{T}^* = T^*$). In this case, T can have several closed extensions, but at most one self-adjoint extension (by maximality). To analyse this situation, one has to consider the spaces

$$D^+ = (T + i)D_T \quad , \quad D^- = (T - i)D_T \quad .$$

(For typographical convenience, we write $T + \lambda$ instead of $T + \lambda 1_H$.)
But

(15.7)   When T is closed and symmetric, $(T + \lambda)D_T$ is closed in H
          for every complex $\lambda \in \mathbb{C}$ with $\text{Im}(\lambda) \neq 0$.

(15.8)   When T is symmetric, $D^+ = D^- = H \iff$ T self-adjoint

In particular, when T is symmetric and both $D^{\pm}$ are dense in H, T is essentially self-adjoint

Coming back to the case T closed and symmetric (hence $D^{\pm}$ closed in H), it is easy to check

$$\| (T + i)x \|^2 = \| Tx \|^2 + \| x \|^2 = \| (T - i)x \|^2 \qquad (x \in D_T)$$

hence there is an isometry $U : D^+ \rightarrow D^-$ which makes the following diagram commutative

$$
\begin{array}{ccc}
 & \xrightarrow{T + i} & D^+ = (T + i) D_T \\
D_T & & \downarrow U \quad \text{isometry} \\
 & \xrightarrow{T - i} & D^-
\end{array}
\qquad .
$$

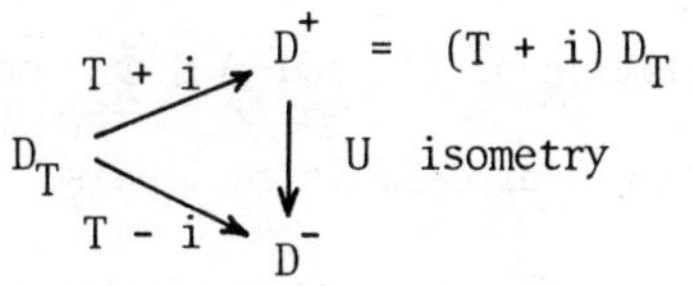

Self-adjoint extensions of T correspond to unitary extensions of U.
Thus we obtain the following criterion. Define the defect indices

$$m^+ = \text{Codim } D^+ \quad , \quad m^- = \text{Codim } D^-$$

for T closed and symmetric (these cardinals can be finite or infinite).

(15.9)   T self-adjoint $\iff$ $(m^+, m^-) = (0,0)$,

(15.10)   T has a self-adjoint extension $\iff$ $(m^+, m^-) = (m,m)$
          ($\iff m^+ = m^- = m$ finite or infinite) .

We can now turn to applications of this theory.

(15.11) <u>Lemma</u>. <u>Let</u> H <u>be a Hilbert space</u>, $\Phi \subset \mathrm{End}(H)$ <u>a (topologically) irreducible set of operators which is invariant under</u> $A \mapsto A^*$ <u>and</u> (T, D_T) <u>a closed (unbounded) operator with</u>

$$A(D_T) \subset D_T \quad \text{for all } A \in \Phi \quad : D_T \text{ is } \Phi\text{-invariant} ,$$

$$AT = TA\big|_{D_T} \quad \text{for all } A \in \Phi .$$

<u>Then</u> $T = \lambda 1_H$ <u>is a scalar operator and</u> $D_T = H$.

<u>Proof</u>. 1) Let us show that D_{T^*} is also Φ-invariant. Take $x \in D_{T^*}$:

$$y \mapsto (x \mid Ty) \quad \text{is continuous on } D_T .$$

As

$$(Ax \mid Ty) = (x \mid A^*Ty) = (x \mid TA^*y)$$

since $A^* \in \Phi$ and $A^*y \in D_T$, the mapping

$$y \mapsto (Ax \mid Ty)$$

can be factorized as $y \mapsto A^*y \mapsto (x \mid TA^*y)$ and is thus continuous: $Ax \in D_{T^*}$ and

$$(x \mid TA^*y) = \begin{cases} (T^*x \mid A^*y) = (AT^*x \mid y) \\ (Ax \mid Ty) = (T^*Ax \mid y) \end{cases} \qquad (y \in D_T)$$

proves the expected commutation relation of T^* and $A \in \Phi$ <u>on</u> D_{T^*} .

2) As $D_{T^*T} = D_T \cap T^{-1} D_{T^*}$, this domain of definition is also invariant under all operators $A \in \Phi$:

$$x \in T^{-1} D_{T^*} \iff Tx \in D_{T^*} \implies ATx \in D_{T^*}$$

furnishes for $x \in D_T$

$$TAx = ATx \in D_{T^*} \quad : \quad Ax \in T^{-1} D_{T^*} \qquad (A \in \Phi).$$

3) The self-adjoint operator T^*T has a spectral decomposition with spectral projectors still commuting to all elements of Φ. By irreducibility assumption of Φ , these spectral projectors can only be 0 or 1_H . But T^*T is a weak limit of linear combinations of spectral projectors . Thus, $T^*T = \lambda 1_H$ must be a scalar operator (and $\lambda \geqslant 0$). We infer

$$\|Tx\|^2 = (Tx \mid Tx) = (T^*Tx \mid x) = \lambda \|x\|^2$$

(first for $x \in D_{T^*T}$ ...). Thus T is continuous and the bounded version of Schur's lemma (8.6) applies: $T = \mu 1_H$ with $|\mu|^2 = \lambda$. q.e.d.

(15.12) Unbounded Schur's lemma. Let (π, H) be a unitary irreducible representation of a group G and (T, D_T) a closed operator with

$$D_T \quad \pi(G)\text{-invariant and } \pi(s)\,T = T\,\pi(s)\big|_{D_T} \qquad (s \in G).$$

Then $T = \lambda\, 1_H$ is a scalar operator (and $D_T = H$).

This is the case $\Phi = \pi(G)$ of (15.11).

From this, we deduce the following statement.

(15.13) Theorem. Let (π, H) and (π', H') be two unitary representations of a group G, $0 \neq T : D_T \rightarrow H'$ a closed operator (D_T dense in H) with

$$D_T \quad \pi(G)\text{-stable}, \quad \pi'(s)\,T = T\,\pi(s)\big|_{D_T} \text{ for all } s \in G .$$

If π is moreover irreducible, T is a multiple of an isometry (hence is continuous with $D_T = H$) and T furnishes an equivalence of π with a subrepresentation of π'.

Proof. One checks, as in part 1 of the proof of (15.11), that D_{T^*} and $D_{T^*T} = D_T \cap T^{-1} D_{T^*} \subset D_T$ are $\pi(G)$-invariant. Since T is closed, T^*T is self-adjoint (in H, and its domain D_{T^*T} is dense). By assumption

$$\pi'(s^{-1})\,T = T\,\pi(s^{-1})\big|_{D_T} \qquad (s \in G)$$

implies

$$T^*\,\pi'(s)\big|_{D_{T^*}} = \pi(s)\,T^*$$

and thus

$$\pi(s)\,T^*Tx = T^*\,\pi'(s)\,Tx = T^*T\,\pi(s)x$$

for $x \in D_{T^*T} \subset D_T$. Thus the unbounded Schur's lemma (15.12) applies to the unbounded operator (T^*T, D_{T^*T}) in H, and the irreducible representation π. Consequently, $T^*T = \lambda\, 1_H$ is continuous and its domain is H (a fortiori $D_T = H$ and T is continuous by the closed graph theorem, but this last fact can be seen independently as we shall see). Thus,

$$(Tx \mid Ty) = (T^*Tx \mid y) = \lambda\,(x \mid y)$$

and T is injective ($T \neq 0 \Longrightarrow \lambda \neq 0$), with $T/\sqrt{\lambda}$ isometric. This implies that the image of T is closed and T (or $T/\sqrt{\lambda}$) an equivalence of π onto $\pi'\big|_{\mathrm{Im}\,T}$. q.e.d.

EXERCISE

Establish (15.11) with the following weaker assumptions. Still assume $\Phi \subset End(H)$ is irreducible in H, (T,D_T) is closed, but replace the commutation of elements $A \in \Phi$ with T <u>on</u> D_T by the following. Let W be a dense subspace of H contained in D_T with

$$W\ \Phi\text{-invariant and } TAx = ATx \text{ for } A \in \Phi \text{ and } x \in W.$$

<u>Hints:</u> a) Define $S = T|_W$ an unbounded operator on $D_S = W$. Thus S has a closed extension (take T) and its closure $\bar{S} = S^{**}$.

b) Let the elements $A \in \Phi$ "act" in $H \times H$ by $A\cdot(x,y) = (Ax,Ay)$. Prove

$$A\cdot Graph(S) \subset Graph(S)$$

hence $A\cdot\overline{Graph(S)} \subset \overline{Graph(S)}$ by continuity.

c) Apply (15.11) to Φ and $(\bar{S},D_{\bar{S}})$.

Formulate the corresponding generalizations of (15.12) and (15.13).

16 DISCRETE SERIES OF LOCALLY COMPACT GROUPS

In this section, G will always denote a locally compact *unimodular* group. Let $\pi : G \longrightarrow Gl(H)$ be a unitary representation of such a group. For each u, $v \in H$, we can form the coefficient c_v^u

$$s \longmapsto (u \mid \pi(s)v) = c_v^u(s) \quad .$$

By definition of the continuity of a representation, these functions are continuous. Since π is unitary, they are also bounded

$$|(u \mid \pi(s)v)| \leqslant \|u\| \|\pi(s)\| \|v\| = \|u\| \|v\| \, .$$

Left and right translates of coefficients are still coefficients

$$c_v^u(x^{-1}s) = (u \mid \pi(x)^* \pi(s)v) = c_v^{\pi(x)u}(s) \, ,$$

$$c_v^u(sx) = (u \mid \pi(s)\,\pi(x)v) = c_{\pi(x)v}^u(s) \quad .$$

Thus, $u \longmapsto c_v^u$ (resp. $v \longmapsto c_v^u$) is a G-morphism from (π, H) to the left (resp. right) regular representation of G in the space of bounded continuous functions.

(16.1) Definition. Let (π, H) be a unitary irreducible representation of G. We say that π is square summable, or equivalently in the discrete series of G when it has one non-zero square summable coefficient

$$0 \neq c_v^u \in L^2(G) \quad \text{for some } u, v \in H \quad .$$

Similarly, we say that π is integrable, when it has a non-zero integrable coefficient

$$0 \neq c_v^u \in L^1(G) \quad \text{for some } u, v \in H \quad .$$

Let us observe that if f is a bounded function, say $|f| \leqslant M$, then $|f|^2 \leqslant M|f|$ so that $f \in L^1 \Longrightarrow f \in L^2$ and

$$\pi \text{ integrable} \Longrightarrow \pi \text{ in the discrete series.}$$

If π is irreducible and $v \neq 0$, the set $\{\pi(s)v : s \in G)\}$ is total in H. Thus we see that $c_v^u \neq 0$ precisely when $u \neq 0$ and $v \neq 0$.

Let us assume now that π is irreducible and in the discrete series with $0 \neq c_v^u \in L^2(G)$. We define

$$W = W^u = \{ w \in H : c_w^u \in L^2(G) \} .$$

This is a linear subspace of H containing all $w = \pi(s)v$, hence <u>dense</u> in H. Moreover, W is invariant under all $\pi(s)$ $(s \in G)$ since the $c^u_{\pi(s)w}$ are right translates of c_w^u. Let us consider the linear operator

$$T : W \to L^2(G) , \quad w \mapsto c_w^u \qquad (T = T^u) ,$$

an *unbounded operator* with domain $D_T = W$. I claim that this operator (T,W) is <u>closed</u>. Let indeed (w_n) be a sequence of elements of W with

$$(w_n , Tw_n) \to (w , f) \quad \text{in} \quad H \times L^2(G) .$$

Thus

$$c^u_{w_n}(s) = (u \,|\, \pi(s)w_n) = (\pi(s^{-1})u \,|\, w_n) \to (\pi(s^{-1})u \,|\, w)$$

and the convergence $c^u_{w_n} \to c^u_w$ is *uniform* on G :

$$\left| c^u_w(s) - c^u_{w_n}(s) \right| = \left| (\pi(s^{-1})u \,|\, w - w_n) \right| \leqslant \|u\| \|w - w_n\| .$$

Extracting a subsequence of (Tw_n), we can assume

$$c^u_{w_n}(s) = Tw_n(s) \to f(s) \quad \text{for} \quad s \notin N \text{ (negligible in G)} .$$

This proves $f(s) = c^u_w(s)$ for $s \notin N$ and thus

$$c^u_w \in L^2(G) , \ w \in W \quad \text{and} \quad f = Tw : \quad (w,f) \in \mathrm{Graph}(T) .$$

Since we are assuming G unimodular, its right regular representation r in $L^2(G)$ is unitary and we can apply (15.13) to T considered as an intertwinning operator between π (on W) and r (on $L^2(G)$). Thus T is a multiple of an isometry and gives an equivalence between π and a subrepresentation of the right regular representation. In particular, $c^u_w \in L^2(G)$ for *all* $w \in W$ and T is *continuous*. Similar considerations would apply to the left regular representation and would show

$$c^u_v \in L^2(G) \text{ for all } u , v \in H .$$

We have thus obtained the essential part of the following statement.

(16.2) <u>Theorem</u>. <u>Let</u> G <u>be a locally compact unimodular group and</u> π <u>a unitary irreducible representation of</u> G. <u>Then the following properties are equivalent</u> :

i) one coefficient $c_v^u \neq 0$ of π is square summable,

ii) all coefficients c_v^u of π are square summable,

iii) π is equivalent to a subrepresentation of the right regular representation.

Proof. We have already proved i) $\Longrightarrow$ ii) $\Longrightarrow$ iii). We show now that iii) $\Longrightarrow$ i). Let (π, H) be a representation which is equivalent to a subrepresentation of the right regular representation in $L^2(G)$. There is no harm in assuming that $H \subset L^2(G)$ and $\pi = r|_H$. We have to show that π has a non-zero square summable coefficient. Take φ and ψ in H with $(\varphi \mid \psi) \neq 0$. Since $C_c(G)$ is dense in $L^2(G)$, the orthogonal projection of $C_c(G)$ on H is also dense in H, and we can assume that $\varphi \in H$ is the orthogonal projection of a $\varphi_o \in C_c(G)$. The corresponding coefficient c_ψ^φ of $\pi = r|_H$ is

$$
\begin{aligned}
c_\psi^\varphi(s) &= (\varphi \mid \pi(s)\psi) = (\varphi_o \mid r(s)\psi) = \\
&= \int \overline{\varphi_o(x)}\, r(s)\psi(x)\, dx = \int \overline{\varphi_o(x)}\, \psi(xs)\, dx = \\
&= \int \varphi_o^*(x)\, \psi(x^{-1}s)\, dx = \varphi_o^* * \psi(s) ,
\end{aligned}
$$

i.e. $$c_\psi^\varphi = \varphi_o^* * \psi = \ell(\varphi_o^*)(\psi) \in L^2(G) .$$

This coefficient is non-zero since $c_\psi^\varphi(e) = (\varphi \mid \psi) \neq 0$. q.e.d.

(16.3) Theorem (Schur's orthogonality relations). Let (π, H) and (π', H') be two unitary irreducible representations in the discrete series of G.

a) There exists a constant $0 < d_\pi \in \mathbb{R}$ such that

$$(c_v^u \mid c_{v'}^{u'}) = \frac{1}{d_\pi} \overline{(u \mid u')}\, (v \mid v') \qquad (u, u', v, v' \in H) ,$$

b) if π and π' are not equivalent,

$$(c_v^u \mid c'^{x}_{y}) = 0 \qquad (u, v \in H,\ x,\ y \in H') .$$

This constant d_π is called formal dimension of π. It depends on the choice of the Haar measure on G : $d'x = c\,dx \Longrightarrow d'_\pi = c^{-1} d_\pi$. But the product $d_\pi dx$ is independent from the choice of Haar measure dx. If G is a compact group, $d_\pi = \dim \pi / m(G)$ so that $d_\pi = \dim \pi$ if the Haar measure is *normalized*.

Proof. Fix u and u' in H and consider the two (continuous) operators

$$\theta, \theta' : H \longrightarrow L^2(G) , \quad v \longmapsto c_v^u \ (\text{resp. } v \longmapsto c_v^{u'}) .$$

Thus we have

$$(c_v^u \mid c_{v'}^{u'}) = (\theta v \mid \theta' v') = (v \mid \theta^* \theta' v')$$

and one checks without difficulty that $\theta^*\theta'$ commutes with π hence is a scalar operator (bounded Schur's lemma (8.6)). There is a constant $a_{u,u'}$ with

$$(c_v^u \mid c_{v'}^{u'}) = a_{u,u'} (v \mid v') \qquad (v\ ,\ v' \in H).$$

One proves similarly

$$(c_v^u \mid c_{v'}^{u'}) = b_{v,v'} \overline{(u \mid u')}$$

(one can either compare the left regular representation and $\check{\pi}$, or transform formally $(c_v^u \mid c_{v'}^{u'}) = (c_{u'}^{v'} \mid c_u^v)$ using the fact that G is unimodular). Hence we can write

$$\frac{\overline{(u \mid u')}}{a_{u,u'}} = \frac{(v \mid v')}{b_{v,v'}} = d_\pi$$

is independent of $u\ ,\ u'\ ,\ v\ ,\ v' \in H$. Taking $u = u'$, $v = v'$ we see that $a_{u,u} > 0$ whence $d_\pi > 0$ and of course

$$(c_v^u \mid c_{v'}^{u'}) = a_{u,u'} (v \mid v') = d_\pi^{-1} \overline{(u \mid u')} (v \mid v') .$$

This proves a). On the other hand, fix $u \in H$ and $x \in H'$ and consider the hermitian form

$$(v\ ,\ y) \mapsto J(v,y) = (c_v^u \mid c'^x_y) .$$

The Cauchy-Schwarz inequality gives

$$|J(v,y)| \leqslant \|c_v^u\| \, \|c'^x_y\| \leqslant C \|v\| \|y\|$$

by the first part a) already proved. Hence there exists a linear operator

$$A : H' \to H$$

with $J(v\ ,\ y) = (v \mid Ay)$ (thus Ay is defined by Riesz' theorem). Taking $v = Ay$

$$\|Ay\|^2 = J(Ay\ ,\ y) \leqslant C \|Ay\| \|y\|$$

$$\Longrightarrow \|Ay\| \leqslant C\|y\| .$$

Thus A is continuous. But one checks without difficulty that A is a G - morphism

$$(v \mid A\pi'(s)y) = J(v,\pi'(s)y) = (c^u_v \mid c'^x_{\pi'(s)y}) =$$
$$= (c^u_v \mid r(s)c'^x_y) = (r(s^{-1})c^u_v \mid c'^x_y) = (c^u_{\pi(s^{-1})v} \mid c'^x_y) =$$
$$= J(\pi(s)^*v,y) = (\pi(s)^*v \mid Ay) = (v \mid \pi(s)Ay) .$$

As π' is not equivalent to a subrepresentation of π, $A = 0$. q.e.d.

(16.4) Proposition. *If a locally compact group G has a finite-dimensional square summable unitary irreducible representation π, it is compact and*

$$d_\pi = \dim \pi / m(G) \qquad (m(G) = \int_G dx) .$$

Proof. Let us take an orthonormal basis (e_i) of the Hilbert space H in which π acts and write

$$1 = \|\pi(s)e_j\|^2 = \sum_i |(e_i \mid \pi(s)e_j)|^2 = \sum_i |c^i_j(s)|^2 .$$

Integrating over G we obtain

$$m(G) = \int_G ds = \int_G \sum_i |c^i_j|^2 ds = \sum_i \|c^i_j\|^2 =$$
$$= \sum_i 1/d_\pi = \dim \pi / d_\pi < \infty .$$

Thus G is compact. q.e.d.

(16.5) Theorem. *Let $\pi \in \hat{G}$ be a square summable unitary irreducible representation of G in H and $f \in L^2(G)$. Then, the integral*

$$\pi(\bar{f}) = \int_G \overline{f(s)}\pi(s)\, ds$$

converges in the weak sense, defines a Hilbert-Schmidt operator with

$$\|\pi(\bar{f})\|_2^2 = \|P_\pi(f)\|^2$$

where $P_\pi : L^2(G) \longrightarrow L^2(G,\pi)$ is the orthogonal projector on the isotypical component of type π in the right regular representation r.

Proof. By definition of a weak integral, for each u and $v \in H$, we have to show that the integral $\int \overline{f(s)}(u \mid \pi(s)v)\, ds$ converges. But

$$(u,v) \mapsto \int \overline{f(s)}\ (u \mid \pi(s)v)\, ds = (f \mid c^u_v)$$

is a continuous hermitian form (it is well defined since f and $c^u_v \in L^2(G)$). By the Cauchy-Schwarz inequality

$$|(f \mid c^u_v)| \leqslant \|f\|_2 \|c^u_v\|_2 \leqslant d_\pi^{-\frac{1}{2}} \|f\|_2 \|u\| \|v\| .$$

This proves that the operator $\pi(\bar{f})$ is defined by $(u \mid \pi(\bar{f})v) = (f \mid c_v^u)$ and is continuous with (uniform) norm

$$\|\pi(\bar{f})\| \leqslant d_\pi^{-\frac{1}{2}} \|f\|_2 \quad .$$

To prove that $\pi(\bar{f})$ is a Hilbert-Schmidt operator, we take an orthonormal basis (e_i) in H . The corresponding coefficients $c_j^i(s) = (e_i \mid \pi(s)e_j)$ make an orthogonal basis of the isotypical component $L^2(G,\pi)$ and $(d_\pi^{\frac{1}{2}} c_j^i)_{i,j}$ is an orthonormal basis of this space. Thus

$$\sum_{i,j} |(e_i \mid \pi(\bar{f})e_j)|^2 = \sum_{i,j} |(f \mid c_j^i)|^2 =$$

$$= \frac{1}{d_\pi} \sum_{i,j} |(f \mid d_\pi^{\frac{1}{2}} c_j^i)|^2 = \frac{1}{d_\pi} \|P_\pi f\|_2^2 \quad .$$

This concludes the proof.

Let us denote by Ω the <u>discrete series</u> of G, namely the set of (classes of) square summable unitary irreducible representations of G. If $\pi, \pi' \in \Omega$ are not equivalent, (16.3) shows that $L^2(G,\pi)$ is orthogonal to $L^2(G,\pi')$. Thus we can consider the closed subspace

$$L_d^2(G) = \widehat{\bigoplus_{\Omega}} L^2(G,\pi) \subset L^2(G)$$

and for $f \in L^2(G)$, we define

$$f_d = \sum_{\Omega} P_\pi(f) = \sum_{\Omega} f_\pi$$

so that in particular

$$\|f_d\|^2 = \sum_{\Omega} \|P_\pi(f)\|^2 = \sum_{\Omega} d_\pi \|\pi(\bar{f})\|_2^2 \quad .$$

We can also write this formula (cf. Ex. 1 at the end of this section)

$$\|f_d\|^2 = \sum_{\Omega} \|P_{\bar{\pi}}(f)\|^2 = \sum d_\pi \|\pi(f)\|_2^2 \quad .$$

Considering the mapping

$$\Omega \ni \pi \mapsto \pi(f)$$

as a *Fourier transform* of f, we see that we have a "Plancherel formula" at least if $f \in L_d^2(G)$. In fact, the restriction to the discrete series $\Omega \subset \hat{G}$ of the *Plancherel measure* is given by the positive masses d_π (the Plancherel measure is dual to the Haar measure and multiplication of dx by a positive constant divides the Plancherel measure by the same constant). Thus Ω appears as the discrete part of the Plancherel measure (cf. sec.21) justifying the terminology adopted for it.

Assume now that the group G contains an *open and compact* subgroup K and normalize the Haar measure of G by $\int_K ds = 1$. Denoting by φ_K the characteristic function of K

$$\varphi_K(s) = 1 \text{ if } s \in K \text{ and } \varphi_K(s) = 0 \text{ if } s \notin K ,$$

we see that

$$\varphi_K^* = \check{\varphi}_K = \varphi_K \quad \text{and} \quad \varphi_K * \varphi_K = \varphi_K .$$

Consequently, if $\pi \in \Omega$ acts in H,

$$\pi(\varphi_K) = P^K : H \longrightarrow H^K$$

is the orthogonal projector on the closed subspace of K-fixed vectors in H. Since $\varphi_K \in L^2(G)$, P^K must be a Hilbert-Schmidt operator by (16.5). Thus $H^K = \mathrm{Im}(P^K)$ must be finite-dimensional and

$$\dim H^K = \| P^K \|_2^2 = \| \pi(\varphi_K) \|_2^2$$

and

$$\sum_{\pi \in \Omega} d_\pi \dim H_\pi^K = \| (\varphi_K)_d \|_2^2 \leqslant \| \varphi_K \|_2^2 = 1 .$$

In particular,

$$d_\pi \dim H_\pi^K \leqslant 1 \quad : \quad \dim H_\pi^K \leqslant 1/d_\pi .$$

(16.6) <u>Proposition</u>. <u>An infinite discrete group</u> $G = \Gamma$ <u>has an empty discrete series</u>.

<u>Proof</u>. We can take $K = \{e\}$ in the preceding computations. Thus, if Γ is discrete and has a square summable (π, H) in the discrete series,

$$H = H^K \implies \dim H = \dim H^K < \infty$$

(by the preceding argument) and Γ is compact (16.4). In this case, Γ must be finite. q.e.d.

(16.7) <u>Proposition</u>. <u>Assume that</u> G <u>contains an open compact subgroup</u> K <u>and a discrete series representation</u> $(\pi, H) \in \Omega$ <u>with</u> $H^K \neq \{0\}$. <u>Then, the normalizer</u> $N_G(K)$ <u>of</u> K <u>in</u> G <u>is compact</u> : $[N_G(K):K] < \infty$.

<u>Proof</u>. Let $n \in N = N_G(K)$. Then $\pi(n)$ leaves H^K invariant

$$\pi(k)\pi(n)v = \pi(n)\pi(n^{-1}kn)v = \pi(n)\pi(k')v = \pi(n)v$$

for all $k \in K$ (hence $k' \in K$). Take an orthonormal basis (e_i) of the finite-dimensional space H^K and write

$$1 = \|\pi(n)e_j\|^2 = \sum_i |c_j^i(n)|^2 .$$

Integrating this equality on the group N (this group is a union of classes of K, hence is open in G) with respect to the restriction of the Haar measure of G (this is a Haar measure of N)

$$\int_N dn = \int_N \sum_i |c_j^i(n)|^2 \, dn \leqslant \sum_i \int_G |c_j^i|^2 \, ds ,$$

hence

$$m(N) \leqslant \sum_i \|c_j^i\|^2 = \sum_i 1/d_\pi = \dim H^K / d_\pi$$

q.e.d.

(Observe that if G is discrete and if we take $K = \{e\}$, $N = G$, so that we obtain a new proof of (16.6)

$$G \text{ discrete}, \ \Omega \neq \emptyset \implies G \text{ finite} .)$$

(16.8) Definition. A Fell group is a locally compact group G for which

$$L^2(G) = L^2_d(G) = \widehat{\bigoplus_{\pi\in\Omega}} L^2(G,\pi) .$$

In particular, compact groups are Fell groups, and generally, Fell groups are the groups having a Plancherel formula

$$\|f\|^2 = \sum_\Omega d_\pi \|\pi(f)\|_2^2$$

valid for all $f \in L^2(G)$. It is interesting to construct *non-compact* Fell groups (recall that in this section, all groups are assumed to be unimodular). Before we give such an example, let us make some general comments when G contains an open compact subgroup K . We denote by $\Omega^K \subset \Omega$ the set of (classes of) discrete series representations (π,H) for which $H^K \neq \{0\}$ (these representations are sometimes called K-spherical).

(16.9) Proposition. If a Fell group G contains an open compact subgroup K

$$\operatorname{Inf}_{\Omega^K} d_\pi > 0 \implies G \text{ compact} .$$

Proof. Since

$$\sum_\Omega d_\pi \dim H_\pi^K = \|(\varphi_K)_d\|^2 \leqslant \|\varphi_K\|^2 = 1$$

(with the Haar measure on G normalized by the condition $m(K) = 1$), our assumption implies that the set Ω^K is finite. But if G is a Fell group,

$$L^2(G) = \widehat{\bigoplus} (H_\pi \widehat{\otimes} H_{\bar\pi}) ,$$

$$L^2(K\backslash G/K) = \bigoplus (H_\pi^K \otimes H_{\bar\pi}^K) \quad (\text{sum over } \Omega^K) .$$

This shows that $L^2(K\backslash G/K)$ is finite-dimensional: G is a finite union of compact cosets KsK. q.e.d.

Let us construct non-compact Fell groups. Let p be a prime and $G_p \subset Gl_2(\mathbb{F}_p)$ be the subgroup consisting of the matrices

$$\begin{pmatrix} a & b \\ 0 & 1 \end{pmatrix} \text{ with } a \in \mathbb{F}_p^\times \text{ and } b \in \mathbb{F}_p .$$

Call K_p the diagonal subgroup (b = 0) of G_p and for any set of primes A, consider the *restricted product*

$$G_A = \prod_{p \in A}{}' G_p \subset \prod_{p \in A} G_p$$

relative to the family $(K_p)_{p \in A}$. By definition, a family $s = (s_p)_A$ of elements $s_p \in G_p$ is in G_A if

$$s_p \in K_p \quad (p \in A) \text{ except for a finite set of } p .$$

Thus $G_A \supset K_A = \prod_{p \in A} K_p$ and the topology of G_A is defined by

G_A induces the product topology on K_A,

K_A is open (and compact) in G_A.

Obviously, G_A is a locally compact group : it is a disjoint union of open cosets of K_A and

$$G_A \text{ compact } \Longleftrightarrow A \text{ finite} .$$

To study representations of G_A, we have to start with the representation theory of the finite groups G_p. Since the derived subgroup (G_p, G_p) of G_p is the unipotent subgroup N_p (defined by a = 1) the abelianized of G_p is isomorphic to $\mathbb{F}_p^\times$ and we can thus construct p-1 characters (1-dimensional irreducible representations) of G_p. As G_p has exactly p distinct conjugacy classes

$$X_o \ (a=1, b=0) , \ X_1 \ (a=1, b\neq 0), \ X_i \ (a=i) \ i = 2,\dots,p-1$$

there is only one other irreducible representation π_p (all this when $p \neq 2$!). The equality

$$\text{Card } G_p = p(p-1) = \underbrace{1 + \dots + 1}_{p-1} + (p-1)^2$$

shows that $\dim \pi_p = p-1$. Since G_p is finite, hence compact and discrete, there are two canonical normalizations of Haar measures on G_p... and we choose the third normalization

$$m_p(K_p) = 1 \quad (\text{hence } m_p(G_p) = p \text{ and } \int_{G_p} = \frac{1}{p-1} \sum_{G_p}) .$$

In an obvious sense, the product $\bigotimes_A m_p = m$ is the Haar measure of G_A normalized by $m(K_A) = 1$.

We also construct a representation π_A of G_A as follows. The left regular representation of G_p in $L^2(G_p/K_p)$ decomposes as

$$1 + \pi_p \quad , \quad \mathbb{C} \oplus V_p \qquad (V_p \text{ orthogonal to the constants})$$

and as G_p is the union of the disjoint cosets K_p , $K_p s K_p$ (where s is the matrix a = b = 1)

$$\dim V_p^{K_p} = 1$$

and we select the unique function $x_p \in V_p$ which is orthogonal to the constants, both left and right K_p-invariant, positive at the neutral element and of norm 1. Relatively to the elements $x_p \in V_p$, we can construct the restricted tensor product

$$H_A = \bigotimes_A (V_p, x_p)$$

of the Hilbert spaces V_p . This space is obtained by completion of the vector space generated by the

$$\otimes y_p \quad , \quad y_p \in V_p \quad \text{and} \quad y_p = x_p \text{ except for finitely many p's.}$$

(The choices of vectors x_p give embeddings of the finite tensor products over finite parts A' of A and H_A is an inductive limit of these finite tensor products.) The unitary representation π_A of G_A in H_A is defined on the total set $\otimes y_p$ by

$$\pi_A(\otimes y_p) = \otimes \pi_p(y_p) \quad .$$

This representation π_A is unitary and irreducible (cf. Ex.3 at the end of this section).

For each subset $B \subset A$, there is a canonical projection $G_A \rightarrow G_B$ (forget the components x_p for $p \notin B$), and the representation π_B can be considered as a unitary irreducible representation of G_A by composition. It is obvious that

$$\dim \pi_B < \infty \iff B \text{ finite } \subset A \quad .$$

By our choice of Haar measure on G_p, the formal dimension d_p of the representation π_p in V_p is

$$d_p = \dim \pi_p / m_p(G_p) = (p-1)/p = 1 - 1/p$$

and similarly, the formal dimension d'_p of the identity representation in dimension 1 of G_p is

$$d'_p = 1/p \quad .$$

Thus we can expect that π_B belongs to the discrete series when

$$d_B = \prod_B (1-1/p) \prod_{A-B} 1/p > 0 \quad .$$

If this is the case, d_B is the formal dimension of π_B.

(16.10) Proposition. Let us assume that the selected subset A of primes satisfies $\sum_A 1/p < \infty$. Then the above constructed group G_A is a Fell group and the discrete series of G_A consists of the unitary irreducible representations π for which the set

$$\{p \in A : \pi|_{G_p} \text{ acts scalarly}\} = S$$

is finite and the formal dimension of such a representation is given by

$$d_\pi = \prod_S 1/p \prod_{A-S} (1-1/p) \quad .$$

The proof of this proposition follows easily from the preceding remarks (cf. also ex.3 below). For more properties of the groups G_A (and other examples of Fell groups), the reader is referred to Robert 1968.

More generally, a good reference for the discrete series is Borel 1972.

EXERCISES

In these exercises, G will still denote a locally compact unimodular group and Ω its discrete series.

1. *Prove* $\pi \in \Omega \Longleftrightarrow \bar{\pi} \in \Omega$, *and* $d_\pi = d_{\bar{\pi}}$ *in this case. Moreover, if H is a Hilbert space with an orthonormal basis and A is a Hilbert-Schmidt operator in H,* $\|A\|_2^2 = \|\bar{A}\|_2^2$. *In particular, if* $A = \pi(f)$, $\bar{A} = \bar{\pi}(\bar{f}) = \overline{\pi(f)}$. *Conclude,*

$$\sum_{\Omega} d_\pi \|\pi(\bar{f})\|_2^2 = \sum_{\Omega} d_\pi \|\pi(f)\|_2^2 \quad .$$

2. *Let* (π, H) *be a unitary irreducible representation of G and assume that* $u, v \in H$ *are such that the corresponding coefficient* $c_v^u \in L^1(G)$ *is integrable. Show that for any unitary irreducible representation* (π', H') *which is not equivalent to* (π, H) *and any coefficient* c'^x_y *(x and* $y \in H'$*) of* π',

$$\int_G \overline{c_v^u(s)}\, c'^x_y(s)\, ds = 0 \quad .$$

3. *With the notations of this section concerning the groups* G_A *(A subset of primes) and their representations* π_B $(B \subset A)$, *prove the following.*

a) *All* π_B *are irreducible (it is enough to show that* π_A *is irreducible, and for this, use a method similar to the one used in (3.4)).*

b) *Show that the coefficient* c_p *of* π_p *with respect to* $u = v = x_p$ *is given by*

$$c_p(s) = 1 \quad \text{for } s \in K_p \ ,$$
$$c_p(s) = -1/(p-1) \quad \text{for } s \notin K_p \ .$$

c) *Define the functions* $c_B = \bigotimes_B c_p \otimes \bigotimes_{A-B} 1$ *and compute their quadratic norm by integrating the functions* $|c_B|^2$ *over the fundamental system of compact (subgroups)*

$$G_{(S)} = \prod_S G_p \times \prod_{p \notin S} K_p \qquad (S \text{ finite in } A) \ .$$

d) *Conclude that* $\pi_B \in \Omega$ *if* $\sum_A 1/p < \infty$ *and* $A - B$ *is finite.*

4. Show that if the discrete series of G is not empty, the center Z of G is compact. (When the center Z of G is not *compact, one can still define discrete series "mod Z" by requiring that one non-zero coefficient - hence all - has an absolute value in $L^2(G/Z)$.)*

17 THE DISCRETE SERIES OF $Sl_2(\mathbb{R})$

Let (π,H) be a unitary irreducible representation in the discrete series of some group G. Fix $u \in H$ with $\|u\|^2 = d_\pi$ and consider the coefficients $c_v = c_v^u$ of π. By definition of the discrete series, they are square summable on G and more precisely (16.3.a) shows that

$$H \longrightarrow L^2(G) \quad , \quad v \longmapsto c_v$$

is an *isometry* which embeds π in the right regular representation r of G. If a sequence $v_n \longrightarrow v$ in H, $c_{v_n} \longrightarrow c_v$ uniformly on G

$$|c_v(s) - c_{v_n}(s)| = |(u \mid \pi(s)(v - v_n))| \leqslant \|u\| \|v - v_n\|.$$

In the model of π consisting of right translations in $\{c_v : v \in H\}$, all functions c_v are continuous and

$$c_{v_n} \longrightarrow c_v \text{ in } L^2(G) \implies c_{v_n} \longrightarrow c_v \text{ uniformly}$$

(because the assumption is equivalent to $v_n \rightarrow v$ in H). This situation is very peculiar and reminds one of properties of holomorphic (or harmonic) functions. Indeed, the discrete series of $Sl_2(\mathbb{R})$ will be constructed in spaces of holomorphic (or anti-holomorphic) functions.

The group $Sl_2(\mathbb{R})$ acts on the upper half-plane $Im(z) > 0$, but it will be more convenient to let it act in the unit disc $|w| < 1$ (conformally equivalent to the upper half-plane) and thus, to use a conjugate SU(1,1) of $Sl_2(\mathbb{R})$ in $Sl_2(\mathbb{C})$. Here is a description of this new group.

The elements $g \in SU(1,1)$ are by definition the matrices

$$g \in Sl_2(\mathbb{C}) \text{ with } g \begin{pmatrix}1 & 0\\ 0 & -1\end{pmatrix} g^* = \begin{pmatrix}1 & 0\\ 0 & -1\end{pmatrix} .$$

Writing $g = \begin{pmatrix}a & b\\ c & d\end{pmatrix}$ we find the conditions

$$\begin{pmatrix}a\bar{a} - b\bar{b} & a\bar{c} - b\bar{d}\\ \bar{a}c - d\bar{b} & c\bar{c} - d\bar{d}\end{pmatrix} = \begin{pmatrix}1 & 0\\ 0 & -1\end{pmatrix} \text{ and } ad - bc = 1 .$$

Thus

$$a\bar{a} - b\bar{b} = 1 \text{ and } \bar{a}c = \bar{b}d$$

and multiplying the second by a

$$a\bar{a}c = \bar{b}ad ,$$

$$\bar{b}ad = c + cb\bar{b} \quad , \quad c = \bar{b}(ad - bc) = \bar{b} \quad .$$

Similarly, $d = \bar{a}$ and thus

(17.1) SU(1,1) consists of the matrices $\begin{pmatrix} a & b \\ \bar{b} & \bar{a} \end{pmatrix}$ with $|a|^2 - |b|^2 = 1$.

This group acts in the unit disc by

$$z \longmapsto \begin{pmatrix} a & b \\ \bar{b} & \bar{a} \end{pmatrix} \cdot z = \frac{az + b}{\bar{b}z + \bar{a}}$$

(observe that if $|z| = 1$, $z = e^{i\varphi}$ and

$$\frac{ae^{i\varphi} + b}{\bar{b}e^{i\varphi} + \bar{a}} = e^{-i\varphi} \frac{ae^{i\varphi} + b}{\bar{a}e^{-i\varphi} + \bar{b}} = e^{-i\varphi} \frac{w}{\bar{w}} .$$

is a complex number of modulus 1). The stabilizer of the origin for this action is obviously defined by

$$b/\bar{a} = 0 \quad , \quad \text{hence } b = 0 \quad .$$

We call T this compact subgroup of SU(1,1) (conjugate to $K \subset Sl_2(\mathbb{R})$)

$$T \ni \begin{pmatrix} e^{i\varphi} & 0 \\ 0 & e^{-i\varphi} \end{pmatrix} \quad .$$

For the convenience of the reader, we give a table of correspondance for this conjugation.

$$\begin{array}{ccc} & Sl_2(\mathbb{C}) & \\ \diagup & & \diagdown \\ Sl_2(\mathbb{R}) & \xrightarrow{\sim} & SU(1,1) \end{array}$$

$$g = \begin{pmatrix} a & b \\ c & d \end{pmatrix} \longmapsto \frac{1}{2i}\begin{pmatrix} 1 & -i \\ 1 & i \end{pmatrix} g \begin{pmatrix} i & i \\ -1 & 1 \end{pmatrix}$$

$$= \begin{pmatrix} \tfrac{1}{2}(a+d)+\tfrac{1}{2}(b-c)i & \tfrac{1}{2}(a-d)-\tfrac{1}{2}(b+c)i \\ \cdots & \cdots \end{pmatrix}$$

$$\frac{1}{2i}\begin{pmatrix} i & i \\ -1 & 1 \end{pmatrix} g \begin{pmatrix} 1 & -i \\ 1 & i \end{pmatrix} \longleftarrow\!\!\shortmid \begin{pmatrix} a & b \\ \bar{b} & \bar{a} \end{pmatrix}$$

$$= \begin{pmatrix} \mathrm{Re}(a+b) & \mathrm{Im}(a-b) \\ -\mathrm{Im}(a+b) & \mathrm{Re}(a-b) \end{pmatrix}$$

$$K \ni \begin{pmatrix} \cos\varphi & \sin\varphi \\ -\sin\varphi & \cos\varphi \end{pmatrix} \longleftrightarrow \begin{pmatrix} e^{i\varphi} & 0 \\ 0 & e^{-i\varphi} \end{pmatrix} \in T$$

$$\begin{pmatrix} e^{t} & 0 \\ 0 & e^{-t} \end{pmatrix} \longleftrightarrow \begin{pmatrix} \mathrm{Ch}\, t & \mathrm{Sh}\, t \\ \mathrm{Sh}\, t & \mathrm{Ch}\, t \end{pmatrix}$$

$$\begin{pmatrix} \mathrm{Ch}\, t & \mathrm{Sh}\, t \\ \mathrm{Sh}\, t & \mathrm{Ch}\, t \end{pmatrix} \longleftrightarrow \begin{pmatrix} \mathrm{Ch}\, t & -i\mathrm{Sh}\, t \\ i\mathrm{Sh}\, t & \mathrm{Ch}\, t \end{pmatrix}$$

Before constructing the discrete series of SU(1,1), we have to give the expression of the invariant measure of the unit disc $X \subset \mathbb{C}$ (isomorphic to SU(1,1)/T). Only an adaptation of the example 5 of sec.12 (p.125) has to be made. For $s = \begin{pmatrix} a & b \\ \bar{b} & \bar{a} \end{pmatrix}$ and $w = s \cdot z$ we have

(17.2) $$\frac{d}{dz} w = (\bar{b}z + \bar{a})^{-2} = j(s,z)^{-2} .$$

The chain rule for derivations gives immediately

$$j(st,z)^2 = j(s,t \cdot z)^2 j(t,z)^2$$

and one can check directly (or extracting roots: $j(s,z) = \bar{b}z + \bar{a}$ does not vanish in the unit disc since $|a|^2 - |b|^2 = 1$ implies $|a| > |b|$)

(17.3) $$j(st,z) = j(s,t \cdot z)\, j(t,z) .$$

On the other hand, one also shows that

$$1 - |w|^2 = |j(s,z)|^{-2} (1 - |z|^2) .$$

In particular, the 2-form on the unit disc X (put $z = x + iy$)

$$(1 - |z|^2)^{-2}\, dz \wedge d\bar{z} = 2i(1 - |z|^2)^{-2}\, dx \wedge dy$$

is invariant under the action of SU(1,1). An invariant measure on X is thus

(17.4) $$m = (1 - r^2)^{-2}\, dx\, dy = (1 - r^2)^{-2}\, r\, dr\, d\varphi \qquad (r = |z|).$$

Let us look at the (quasi-invariant) measures

$$m_k = (1 - r^2)^k\, m = (1 - r^2)^{k-2}\, dx\, dy$$

on X and form the Hilbert spaces $H_k = L^2(X, m_k)$. The group G = SU(1,1) acts in a natural way in H_k by

(17.5) $$\pi_k(s)f\,(z) = j({}^t s,z)^{-k}\, f({}^t s \cdot z) .$$

The cocycle condition (17.3) shows immediately that $\pi_k(st) = \pi_k(s)\, \pi_k(t)$. Moreover, since $j(s,z) \to 1$ uniformly on compact sets when $s \to e$, the continuity of $s \mapsto \pi_k(s)f$ follows (sec.13 p.132) and π_k is a representation of G in H_k .

(17.6) Lemma. The representation (π_k, H_k) of G = SU(1,1) is unitary.

Proof. By definition,

$$\|\pi_k(s)f\|^2 = \int |j({}^t s,z)|^{-2k} |f({}^t s \cdot z)|^2 (1 - |z|^2)^k\, dm(z) .$$

Make the change of variable $z' = {}^t s\cdot z$ and use the invariance of m

$$\|\pi_k(s)f\|^2 = \int |j({}^t s,\check{s}\cdot z)|^{-2k}(1-|\check{s}\cdot z|^2)^k |f({}^t s\cdot z)|^2\, dm(z) =$$

$$= \int \left|j({}^t s,\check{s}\cdot z)j(\check{s},z)\right|^{-2k} |f(z')|^2 (1-|z'|^2)^k\, dm(z') = \|f\|^2 .$$

Thus the lemma follows.

The representations π_k in H_k are *reducible*. Indeed, let us consider the vector subspace H_k^{hol} consisting of holomorphic functions in $H_k = L^2(X,m_k)$.

(17.7) Proposition. The subspace H_k^{hol} is closed in H_k and

$$H_k^{hol} \neq \{0\} \text{ for } k \geqslant 2 .$$

Proof. The second assertion is obvious since all polynomials in z belong to H_k when $k \geqslant 2$ ($m_k = (1-r^2)^{k-2}\,dx\,dy$ and $k-2 \geqslant 0$). To prove the first one, we use the mean value property of holomorphic functions as follows. Let f be holomorphic on a disc $|z-a| \leqslant r$:

$$\iint_{|z-a|<r} f(z)\,dx\,dy = \int_0^r \rho\,d\rho \int_0^{2\pi} f(a+\rho e^{i\theta})\,d\theta =$$

$$= 2\pi \int_0^r \rho\,d\rho \cdot f(a) = \pi r^2 f(a) .$$

The preceding computation is valid for any holomorphic f on the disc X provided $|a| \leqslant 1-r$ (say $0 < r < 1$). Thus,

$$\pi r^2 |f(a)| \leqslant \iint_{|z-a|<r} |f(z)|\,dx\,dy =$$

$$= \iint_{|z-a|<r} \left|f(z)(1-r^2)^{\frac{1}{2}(k-2)}\right| (1-r^2)^{\frac{1}{2}(2-k)}\,dx\,dy$$

$$\leqslant \left(\iint_{|z-a|<r} |f(z)|^2(1-r^2)^{k-2}\,dx\,dy\right)^{1/2} \left(\iint_{|z-a|<r} (1-r^2)^{2-k}\,dx\,dy\right)^{1/2} \leqslant$$

$$\leqslant \|f\| \left(\iint_{|z-a|<r} (1-r^2)^{2-k}\,dx\,dy\right)^{1/2} .$$

Let us take $0 < r < \frac{1}{2}$ and $|a| < 1-2r$ so that the disc $|z-a| \leqslant r$ is contained in the fixed disc $|z| \leqslant 1-r$. Then the integral of $(1-r^2)^{2-k}$ over this disc is finite and we can write

$$|f(a)| \leqslant C_r \|f\| \qquad (|a| \leqslant 1-2r).$$

This shows that convergence in L^2 implies uniform convergence on all compact subsets of X for sequences of H_k^{hol}. Take now a convergent sequence

$$(f_n) \subset H_k^{hol} \ , \quad f_n \longrightarrow f \ \text{ in } \ H_k = L^2(X, m_k) \ .$$

Thus (f_n) must also be a Cauchy sequence for the uniform convergence on compact subsets of X and it has a limit g, which is a holomorphic function on X. But extracting if necessary a subsequence of (f_n), we can assume that $f_n(z) \longrightarrow f(z)$ for $z \notin N$ (m_k-negligible). In particular, this shows $f(z) = g(z)$ for $z \notin N$, hence $f = g \in H_k$ q.e.d.

(17.8) Theorem. For $k \geqslant 2$, the unitary representations (π_k, H_k^{hol}) are irreducible, in the discrete series of $G = SU(1,1)$ and the formal dimension d_k of π_k is $d_k = (k-1)/\pi$ for the choice of Haar measure on G given by

$$\int_G f(s)\, ds = \int_X dm(z) \int_T dt\, f(st) \qquad (\ z = s\cdot 0\ )$$

where dt is the normalized Haar measure of the compact group T.

Proof. The proof of (17.7) has shown that the evaluation forms

$$\varepsilon_z : H_k^{hol} \longrightarrow \mathbb{C} \ , \quad f \longmapsto f(z)$$

are continuous. By the Riesz theorem, they are given by scalar products with some vectors $e_z \in H_k^{hol}$:

$$e_z \in H_k^{hol} \text{ is defined by } f(z) = (e_z \mid f) \text{ for all } f \in H_k^{hol} \ .$$

The set $\{e_z : z \in X\}$ is *total* in H_k^{hol} :

$$(e_z \mid f) = 0 \ \text{ for all } z \in X \implies f(z) = 0 \ \text{ for all } z \in X$$
$$\implies f = 0 \in H_k^{hol} \ .$$

By definition of π_k we have

$$\pi_k({}^t s)\, e_\zeta(z) = j(s,z)^{-k}\, e_\zeta(sz)$$

whence

$$(e_z \mid \pi_k({}^t s)\, e_\zeta) = j(s,z)^{-k}\, (e_{sz} \mid e_\zeta)$$

and

$$(\pi_k(\check{s})\, e_z \mid e_\zeta) = (\overline{j(s,z)}^{-k}\, e_{sz} \mid e_\zeta)$$

for all $\zeta \in X$. By the totality of the set of e_ζ, this implies

$$\pi_k(\check{s})\, e_z = \overline{j(s,z)}^{-k}\, e_{sz} \ .$$

Thus, for z and ζ in X

$$
\begin{aligned}
(e_{s\zeta} \mid e_{sz}) &= \left(\overline{j(s,\zeta)}^k \pi_k(\check{s})\, e_\zeta \,\middle|\, \overline{j(s,z)}^k \pi_k(\check{s})\, e_z \right) \\
&= j(s,\zeta)^k\, \overline{j(s,z)}^k\, (\pi_k(\check{s}) e_\zeta \mid \pi_k(\check{s}) e_z) \\
&= j(s,\zeta)^k\, \overline{j(s,z)}^k\, (e_\zeta \mid e_z)
\end{aligned}
$$

since $\pi_k(\check{s})$ is a unitary operator. Now, for any operator $A \in \mathrm{End}(H_k^{hol})$ which commutes to all $\pi_k(s)$ $(s \in G)$ we can check that

$$(e_\zeta \mid Ae_z) \quad \text{and} \quad (e_\zeta \mid e_z)$$

have the same transformation law under the substitution $(\zeta, z) \mapsto (s\zeta, sz)$. This proves that the function $h = h_A$ defined by

$$h(\zeta, z) = (e_\zeta \mid Ae_z) \,/\, (e_\zeta \mid e_z)$$

is invariant :

$$h(s\zeta, sz) = h(\zeta, z) \qquad (s \quad G) .$$

But $(e_\zeta \mid e_z) = e_z(\zeta)$ is holomorphic in ζ (and anti-holomorphic in z by permutation of z and ζ). Thus h is meromorphic in ζ. Take $\zeta \neq 0$ and $t \in T$ (stabilizer of 0) :

$$h(\zeta, 0) = h(t\cdot\zeta, t\cdot 0) = h(t\cdot\zeta, 0) .$$

The points $t\cdot\zeta$ describe a circle centered at 0 (with radius $|\zeta| \neq 0$). Since h is meromorphic, it must be constant : $h(\zeta, 0) = c$. Now, for each $z \in X$, there exists a $s_z \in G$ with $s_z \cdot z = 0$ whence

$$h(\zeta, z) = h(s_z\zeta, 0) = c .$$

Consequently, h is constant (= c) and

$$(e_\zeta \mid Ae_z) = c\,(e_\zeta \mid e_z) = (e_\zeta \mid c\, e_z) \qquad (\zeta, z \in X) .$$

This proves $Ae_z = c\,e_z$ for all $z \in X$ and $A = c\,1$ is a scalar operator. The only projectors which commute to π_k on H_k^{hol} are 0 and 1 : (π_k, H_k^{hol}) is <u>irreducible</u>. It is easy to determine a coefficient of this representation. We take $u = v$ to be the constant function $1 \in H_k^{hol}$ and compute $c = c_1^1 = c_v^u$:

$$c_u^u(s) = \iint \overline{u(z)}\, j({}^t s, z)^{-k}\, u({}^t s\cdot z)\, (1 - |z|^2)^{k-2}\, dx\, dy ,$$

$$c_1^1(s) = \iint (bz + \bar{a})^{-k}\, (1 - |z|^2)^{k-2}\, dx\, dy \quad \text{if} \quad s = \begin{pmatrix} a & b \\ \bar{b} & \bar{a} \end{pmatrix} .$$

Using polar coordinates, we have

$$c_1^1(s) = \int_0^1 dr\, r\,(1 - r^2)^{k-2} \int_0^{2\pi} (\bar{a} + b\, r\, e^{i\theta})^{-k}\, d\theta \quad .$$

But the integral $\int_0^{2\pi} \frac{d\theta}{(\alpha + \beta e^{i\theta})^k}$ is easily computed with the method of residues :

$$\text{let us put} \quad z = e^{i\theta} \;,\; dz = ie^{i\theta} d\theta \;,\; d\theta = (iz)^{-1} dz$$

$$\int_0^{2\pi} \ldots = \oint_{|z| = 1} (iz)^{-1}(\alpha + \beta z)^{-k} dz = 2\pi i \;\text{Residue}_{z=0}$$

$$= 2\pi/\alpha^k \quad \text{when} \quad |\alpha| > |\beta| .$$

In our case,

$$c_1^1(s) = c_1^1(a,b) = \frac{2\pi}{\bar{a}^k} \int_0^1 dr\, r\,(1 - r^2)^{k-2} = \frac{\pi}{(k-1)\,\bar{a}^k} \quad .$$

From here, it is easy to compute the quadratic norm of this function on G. Since u = v = 1 is T invariant, the function $c_v^u = c_1^1$ is (left and) right T-invariant and the integral over G reduces to an integral over X (we use the normalization of ds on G, with the normalization of dt on T). We have to consider this function c_1^1 on G as a function on X via the canonical mapping

$$G \to G/T \xrightarrow{\sim} X \;,\; s \mapsto s\cdot 0 = z \quad .$$

But if $s = \begin{pmatrix} a & b \\ \bar{b} & \bar{a} \end{pmatrix}$, $s\cdot 0 = b/\bar{a}$ and

$$|a|^2 |z|^2 = |b|^2 = |a|^2 - 1 \;,$$

$$|a|^2(1 - |z|^2) = 1 \;,\quad 1 - |z|^2 = |a|^{-2} \quad .$$

This gives

$$\|c_1^1\|^2 = \iint_X \frac{\pi^2}{(k-1)^2} |a|^{-2k}\, dm(z) = \frac{\pi^2}{(k-1)^2} \iint (1 - r^2)^{k-2}\, r\, dr\, d\theta$$

$$= \frac{\pi^2}{(k-1)^2}\, 2\pi \frac{1}{2}\, \frac{1}{k-1} = \frac{\pi^3}{(k-1)^3} \quad .$$

But it is easy to check that the norm of $1 \in H_k^{hol}$ is given by $\|1\|^2 = \frac{\pi}{k-1}$.
We have thus proved

$$\|c_1^1\|^2 = \pi \|1\|^4 \frac{1}{k-1} \quad .$$

The formal dimension of (π_k, H_k^{hol}) is thus $d_k = \frac{1}{\pi}(k-1)$. q.e.d.

EXERCISES

1. For $k \geqslant 2$, give the decomposition of the restriction of (π_k, H_k^{hol}) to the compact subgroup T (circle group). (For $s = \begin{pmatrix} a & 0 \\ 0 & \bar{a} \end{pmatrix}$, $a\bar{a} = 1$, show that

$$\pi_k(s)\; z^{\ell} \;=\; a^{k+2\ell}\; z^{\ell} \;.)$$

2. Using anti-holomorphic functions in H_k ($k \geqslant 2$), construct representations in the discrete series of $SU(1,1)$ which are inequivalent to the representations constructed in (17.8). (Calling π_{-k} the representation constructed with antiholomorphic functions, show that the restrictions to T of π_k and π_{-k} are inequivalent: use Ex.1.)

3. Show that (π_k, H_k^{hol}) is integrable when $k \geqslant 3$.

4. Let $e_j(z) = z^{j-k}/\|z^{j-k}\|$, so that $(e_j)_{j \geqslant k}$ is an orthonormal basis of the representation (π_k, H_k^{hol}).

a) Compute the norm

$$\|z^j\|^2 \;=\; \pi\,\frac{j!\,(m-2)!}{(m+j-1)!} \;=\; \frac{\pi}{m-1}\Big/\binom{j+m-1}{j} \;.$$

b) Prove that (notations of the proof of (17.8))

$$e_\zeta(z) \;=\; \sum_{j \geqslant k} \overline{e_j(\zeta)}\, e_j(z) \;=\; \sum_{j \geqslant k} \|z^j\|^{-2}\, \bar{\zeta}^j\, z^j \;.$$

In particular, $e_\zeta(z)$ is not a polynomial in z when $\zeta \neq 0$ ($e_0(z) = \frac{m-1}{\pi}$ is a multiple of 1).

c) Compute $e_{s\cdot 0}(z)$ for $s = \begin{pmatrix} a & b \\ \bar{b} & \bar{a} \end{pmatrix} \in SU(1,1)$.

18 THE PRINCIPAL SERIES OF $Sl_2(\mathbb{R})$

We shall need the following general result.

(18.1) Proposition. Let X be a locally compact measured space and A a bounded operator in $L^2(X)$ which commutes to all multiplication operators

$$M_\varphi : f \mapsto \varphi f \qquad (\varphi \in L^\infty(X)).$$

Then A is itself a multiplication operator $A = M_\psi$ for a certain $\psi \in L^\infty(X)$.

This proposition states that the commutative algebra of multiplication operators is a maximal abelian (von Neumann) subalgebra of $\mathrm{End}(L^2(X))$.

Proof. For each compact set $K \subset X$, denote by φ_K the characteristic function of K and put

$$A(\varphi_K) = \psi_K \in L^2(X) \quad .$$

For $K \subset K'$, we have $\varphi_K = \varphi_K \varphi_{K'}$ whence

$$A(\varphi_K) = A(\varphi_K \varphi_{K'}) = \varphi_K A(\varphi_{K'})$$

since by assumption, A commutes with the multiplication operator by $\varphi_K \in L^\infty(X)$. From this identity $\psi_K = \varphi_K \psi_{K'}$, we infer that

$$\psi_{K'} \text{ (restricted to } K') \text{ extends } \psi_K|_K \quad .$$

Hence there exists a function ψ with

$$\psi|_K = \text{restriction of } \psi_K \text{ to } K$$

for all compact subsets K of X. Take now $f \in C_c(X)$ and call K its support. We must have

$$A(f) = A(\varphi_K f) = f\, A(\varphi_K) = f\, \psi_K = \psi f \quad .$$

Moreover, as is well known,

$$\operatorname*{Sup\,ess}_{x \in K} |\psi(x)| \leqslant \|A\|$$

and $\psi \in L^\infty(X)$. The continuous operator A must coïncide with M_ψ since it coïncides with M_ψ on the dense subspace $C_c(X)$ of $L^2(X)$. q.e.d.

Let us come back to $G = Sl_2(\mathbb{R})$. We introduce its subgroups

K (circle group) consisting of the matrices

$$\begin{pmatrix} \cos\theta & -\sin\theta \\ \sin\theta & \cos\theta \end{pmatrix} \qquad (0 \leq \theta < 2\pi),$$

A diagonal subgroup consisting of matrices $\begin{pmatrix} a & 0 \\ 0 & 1/a \end{pmatrix}$ $(a \in \mathbb{R}^{\times})$,

N unipotent subgroup consisting of matrices

$$\begin{pmatrix} 1 & n \\ 0 & 1 \end{pmatrix} \qquad (n \in \mathbb{R}) ,$$

$P = AN = NA$ upper triangular subgroup (semi-direct product).

Thus we have

$$G = KAN = KP = PK = ANK = NAK = \ldots$$

The product formula $G = NAK$ can be proved for instance as follows. The group G acts transitively on the upper half-plane $\mathrm{Im}\, z > 0$ $(z \in \mathbb{C})$ by fractional linear transformations : the subgroup NA already acts transitively as the following formula shows

$$\begin{pmatrix} 1 & b \\ 0 & 1 \end{pmatrix}\begin{pmatrix} a & 0 \\ 0 & 1/a \end{pmatrix}\cdot i = a^2 i + b .$$

The stabilizer of the point i is K. Thus if $g \in G$ is an arbitrary element, we can put $z = g\cdot i$ and find $n \in N$, $a \in A$ with $na\cdot i = z$. Thus

$$g\cdot i = na\cdot i \implies (na)^{-1}g \in \mathrm{Stab}(i) = K$$

and we have found a decomposition $g = nak$ with $k = (na)^{-1}g$.

As we have seen in sec.13 (example 2, p.123),

$x^{-2}\,dx\,dy$ is a Haar measure on $P \ni \begin{pmatrix} x & y \\ 0 & 1/x \end{pmatrix}$,

and the modular function on P is

$$\Delta_P\begin{pmatrix} x & y \\ 0 & 1/x \end{pmatrix} = 1/x^2 .$$

Let us consider the character α of A defined by $\alpha\begin{pmatrix} a & 0 \\ 0 & 1/a \end{pmatrix} = a^2$ so that

$$\Delta_P(an) = \alpha(a)^{-1} \qquad (a \in A ,\, n \in N).$$

We extend the character trivially on N in order to be able to write

$$\Delta_P(p) = \alpha(p)^{-1} \qquad (p \in P) .$$

The representations of the principal series of $Sl_2(\mathbb{R})$ will now be constructed by an induction procedure from characters of A (compare with sec.8). For the sake of simplicity, we only consider the characters

α^s (leaving as an exercise the consideration of the characters $\alpha^s \cdot \mathrm{sgn}(a)$). The representation π_s acts by right translations in the Hilbert space H_s of (classes of) functions over G satisfying

$$(18.2) \quad \begin{cases} f|_K \in L^2(K) : \|f\|_s^2 = \|f\|_K^2 = \int_K |f(k)|^2 \, dk \\ f(n\begin{pmatrix} a & 0 \\ 0 & 1/a \end{pmatrix} g) = \alpha\begin{pmatrix} a & 0 \\ 0 & 1/a \end{pmatrix}^s f(g) = a^{2s} f(g) . \end{cases}$$

(18.3) Lemma. When $\mathrm{Re}(s) = \frac{1}{2}$, the representation (π_s, H_s) is unitary.

Proof. Since K is compact, G/K has a G-invariant measure: the quotient mapping $G \to G/K$ is proper, so that by composition it induces a transpose $C_c(G/K) \to C_c(G)$ and the image of the Haar measure ds of G is such a measure on G/K. By definition,

$$\int_G f(s)\, ds = \int_{G/K} d\dot{s} \int_K f(sk)\, dk \qquad (f \in C_c(G)).$$

This measure $d\dot{s}$ on G/K must in particular be invariant under left translations by elements of P, and taking for each element of G/K its representative in P, we see that we can identify $d\dot{s}$ with dp (Haar measure of P) and write

$$\int_G f(s)\, ds = \int_P dp \int_K f(pk)\, dk .$$

We can now prove the lemma. We have to evaluate the quadratic norm of a restriction to $K : k \mapsto f(kg)$ of a right translate of an $f \in H_s$. For this, we write $kg = p'_k \cdot k'$ (where p'_k and k' depend on g, but this element is fixed throughout this proof so that only the dependence in k is essential and visible). Thus

$$\|\pi_s(g) f\|_K^2 = \int_K |f(kg)|^2 \, dk = \int_K |f(p'_k k')|^2 dk .$$

By definition of H_s , we can write

$$f(p'_k k') = \Delta_P(p'_k)^{-s} f(k') ,$$

$$|f(p'_k k')|^2 = \Delta_P(p'_k)^{-2\mathrm{Re}(s)} |f(k')|^2 =$$

$$= \Delta_P(p'_k)^{-1} |f(k')|^2 \quad \text{if } \mathrm{Re}(s) = \tfrac{1}{2} .$$

Choose $\varphi \in C_c(P)$ with $\int_P dp\, \varphi(p) = 1$ and assume $f|_K$ continuous. Put

$$\psi = |f|^2 \in C(K) , \quad F(pk) = \varphi(p)\, \psi(k) ,$$

so that $F = \varphi \otimes \psi \in C_c(PK) = C_c(G)$. Thus

$$\| \pi_s(g) f \|^2 = \int_K \Delta_P^{-1}(p'_k)\, \psi(k')\, dk \int_P \varphi(p)\, dp =$$

$$= \int_{P \times K} \varphi(p)\, \psi(k')\, \Delta_P^{-1}(p'_k)\, dp\, dk = \int_K dk \int_P dp\, F(pk')\, \Delta_P^{-1}(p'_k) =$$

$$= \int_K dk \int_P dp\, F(pp'_k k') = \int_K dk \int_P dp\, F(pkg) =$$

$$= \int_G ds\, F(sg) = \int_G ds\, F(s)$$

since G is unimodular ! We have thus found

$$\| \pi_s(g) f \|_K^2 = \int_G F(s)\, ds = \int_P dp \int_K dk\, F(pk) =$$

$$= \int_P dp\, \varphi(p) \int_K dk\, \psi(k) = \int_K dk\, |f(k)|^2 = \|f\|_K^2 .$$

The lemma follows by density of the space $C(K)$ in $L^2(K)$. q.e.d.

(18.4) Theorem. The unitary representations $(\pi_s, H_s)_{\mathrm{Re}(s) = \frac{1}{2}}$ of the group $Sl_2(\mathbb{R})$ constructed above are irreducible when $s \neq \frac{1}{2}$.

Proof. This irreducibility is best seen on another model. Thus we explain how this other model is obtained from H_s. If

$$k_\theta = \begin{pmatrix} \cos\theta & -\sin\theta \\ \sin\theta & \cos\theta \end{pmatrix} = \begin{pmatrix} \alpha & * \\ 0 & 1/\alpha \end{pmatrix} \begin{pmatrix} 1 & 0 \\ x & 1 \end{pmatrix}$$

we have $\alpha = 1/\cos\theta$, $x = \mathrm{tg}\,\theta$ (the representations π_s are trivial on the element $-1 \in Sl_2(\mathbb{R})$, so that we can consider them as representations of $PSl_2(\mathbb{R}) = Sl_2(\mathbb{R})/(\pm 1)$ and thus take $\theta \in [-\frac{1}{2}\pi, \frac{1}{2}\pi[$). Thus

$$f(k_\theta) = f\left(\begin{pmatrix} 1/\cos\theta & * \\ 0 & \cos\theta \end{pmatrix} \begin{pmatrix} 1 & 0 \\ \mathrm{tg}\,\theta & 1 \end{pmatrix}\right) =$$

$$= (\cos\theta)^{-2s} f\begin{pmatrix} 1 & 0 \\ \mathrm{tg}\,\theta & 1 \end{pmatrix} .$$

Take $x = \mathrm{tg}\,\theta$ as a new variable :

$$dx = (\mathrm{tg}^2\theta + 1)\, d\theta \quad , \quad d\theta = dx/(x^2 + 1) ,$$

and put

$$\varphi(x) = \varphi_f(x) = f\begin{pmatrix} 1 & 0 \\ x & 1 \end{pmatrix} .$$

We have

$$\infty > \int_K |f(k)|^2\, dk = \frac{1}{\pi} \int_{-\pi/2}^{\pi/2} |f(k_\theta)|^2\, d\theta =$$

$$= \frac{1}{\pi}\int_{-\infty}^{\infty} |(x^2+1)^s \varphi(x)|^2 (x^2+1)^{-1}\,dx =$$

$$= \frac{1}{\pi}\int_{-\infty}^{\infty} |\varphi(x)|^2\,dx \qquad \text{(since } \mathrm{Re}(s) = \tfrac{1}{2}).$$

Thus

$$f \mapsto \varphi = \varphi_f$$

defines an isometry $H_s \xrightarrow{\sim} L^2(\mathbb{R},dx)$. Let us determine the action of π_s in this new model :

$$\begin{pmatrix}1&0\\x&1\end{pmatrix} g = \begin{pmatrix}1&0\\x&1\end{pmatrix}\begin{pmatrix}a&b\\c&d\end{pmatrix} = \begin{pmatrix}a&b\\ax+c&bx+d\end{pmatrix} = \begin{pmatrix}\alpha&\beta\\0&1/\alpha\end{pmatrix}\begin{pmatrix}1&0\\y&1\end{pmatrix}$$

$$\Longrightarrow \quad 1/\alpha = bx+d \quad \text{and} \quad y/\alpha = ax+c$$

$$\Longrightarrow \quad y = \frac{ax+c}{bx+d} \quad \text{and} \quad \alpha^2 = (bx+d)^{-2} .$$

As we have

$$\pi_s(g)\, f\begin{pmatrix}1&0\\x&1\end{pmatrix} = f\left(\begin{pmatrix}1&0\\x&1\end{pmatrix} g\right) = f\left(\begin{pmatrix}\alpha&\beta\\0&1/\alpha\end{pmatrix}\begin{pmatrix}1&0\\y&1\end{pmatrix}\right) =$$

$$= \alpha^{2s} f\begin{pmatrix}1&0\\y&1\end{pmatrix} ,$$

the preceding formulas show that

$$[\pi_s(g)\varphi](x) = (bx+d)^{-2s}\, \varphi\left(\frac{ax+c}{bx+d}\right) .$$

In particular, the operators $\pi_s\begin{pmatrix}1&0\\c&1\end{pmatrix}$ are *translations* and the operators $\pi_s\begin{pmatrix}a&0\\0&1/a\end{pmatrix}$ are suitably normalized *dilatations* :

$$\pi_s\begin{pmatrix}a&0\\0&1/a\end{pmatrix}\varphi(x) = a^{2s}\,\varphi(a^2x) ,$$

$$|a^{2s}\varphi(a^2x)|^2 = a^2|\varphi(a^2x)|^2 \qquad \text{(since } \mathrm{Re}(s) = \tfrac{1}{2})$$

and thus

$$\left|\pi_s\begin{pmatrix}a&0\\0&1/a\end{pmatrix}\varphi\right|^2 dx = a^2|\varphi(a^2x)|^2\,dx = |\varphi|^2\,dx .$$

Let us now effect a Fourier transform

$$\hat{\varphi}(\xi) = \int_{-\infty}^{\infty}\varphi(x)\, e^{-ix\xi}\,dx .$$

The operators $\pi_s\begin{pmatrix}1&0\\c&1\end{pmatrix}^{\wedge}$ will now be multiplication operators by the characters

$$\chi_c : \xi \mapsto e^{ic\xi}$$

($|\chi| = 1$ hence M_χ is a unitary operator in $L^2(\mathbb{R},d\xi)$).

The announced irreducibility will now follow from Schur's lemma. Any continuous operator which commutes with all $\pi_s(g)$ will have to commute with all multiplication operators M_χ (χ unitary character of $\mathbb{R}$) in $L^2(\mathbb{R},d\xi)$. But, for any locally compact measured space X, the mapping $L^\infty(X) \longrightarrow \mathrm{End}(L^2(X))$, $\varphi \longmapsto M_\varphi$ is continuous when $L^\infty(X)$ has the weak topology coming from the duality with $L^1(X)$ and $\mathrm{End}(L^2(X))$ has the weak operator topology. Since this weak operator topology is the weakest for which all mappings

$$A \longmapsto (\theta \mid A\psi) \qquad\qquad (\theta\ ,\ \psi \in L^2(X)\ )$$

are continuous, it is enough to check that the composites

$$\varphi \longmapsto (\theta \mid M_\varphi \psi) = (\theta \mid \varphi\psi) = \int \bar{\theta}\psi \cdot \varphi \, dm(x)$$

are continuous on $L^\infty(X)$ (equipped with its weak topology). But since

$$\theta\ ,\ \psi \in L^2(X) \implies \bar{\theta}\psi \in L^1(X) \quad ,$$

the preceding assertion follows by definition. We infer from this that if an operator T commutes with all multiplication operators M_χ, it will still commute with all multiplication operators M_φ, φ being in the (weak) closure of the space generated by the $\chi \in L^\infty(X)$ (composition of operators is separately continuous for the weak operator topology). Thus (18.1) will be applicable as soon as we have proved that for any locally compact non-discrete field F, the set of additive unitary characters $\chi \in \hat{F}$ is *total* in $L^\infty(F)$ for the weak topology defined by its duality with $L^1(F)$. By the Hahn-Banach theorem, it is enough to prove that

$$f \in L^1(F)\ ,\ \forall \chi \in \hat{F}\ ,\quad \langle f,\chi\rangle = 0 \implies f = 0\ .$$

But this is a classical result of the theory of Fourier transforms since

$$\langle f,\chi\rangle = \int f(x)\,\chi(x)\,dx = \hat{f}\,(\chi^{-1}) \quad :$$

indeed

$$f \in L^1(F)\ ,\ \hat{f} = 0 \implies f = 0\ .$$

From (18.1), we now infer that the operator T which commutes with the representation π_s in $L^2(\mathbb{R},d\xi)$ is a multiplication operator M_φ by a certain function $\varphi \in L^\infty(\mathbb{R},d\xi)$. Since T also commutes with the dilatations, φ must be invariant under dilatation

$$\int_{-\infty}^{\infty} a^{2s}\,\psi(a^2x)\,e^{-ix\xi}\,dx = \int_{-\infty}^{\infty}\psi(a^2x)\,e^{-ia^2x\,\xi/a^2}\,a^{2s-2}d(a^2x) =$$

$$= a^{2(s-1)}\,\hat{\psi}\,(\xi/a^2) \quad ,$$

and the commutation relation implies

$$a^{2(s-1)}(\hat{\psi}\,\varphi)\,(\xi/a^2) = a^{2(s-1)}\,\hat{\psi}\,(\xi/a^2)\,\varphi(\xi)\ \text{for all}\ \hat{\psi}\in L^2 \ .$$

This implies precisely that φ is invariant under positive dilatations

$$\varphi(\xi) = \mu\,\mathrm{sgn}\,\xi + \lambda \quad .$$

One can see that for $s \neq \frac{1}{2}$, only the functions with $\mu = 0$ furnish operators commuting with π_s : $T = M_\varphi = \lambda\cdot 1$. But for $s = \frac{1}{2}$, the multiplication operator corresponding to

$$Y(\xi) = \tfrac{1}{2}(\mathrm{sgn}\,\xi + 1)$$

is a non-trivial projector commuting to $\pi_{\frac{1}{2}}$ and

$$\pi_{\frac{1}{2}} = \pi^+_{\frac{1}{2}} \oplus \pi^-_{\frac{1}{2}}$$

is a decomposition into irreducible components . q.e.d.

19 DECOMPOSITION ALONG A COMMUTATIVE SUBGROUP

Let G be a locally compact group and A a closed (hence locally compact) commutative subgroup. In order to have an invariant measure on the homogeneous space G/A, we assume that $\Delta_G|_A = 1$ (cf.(12.8)). We shall use the duality theory for locally compact abelian groups to give a decomposition of the left regular representation of G in $L^2(G)$.

Let $\hat{A}$ be the Pontryagin dual of A : it is a locally compact abelian group having for elements the unitary characters $\chi : A \longrightarrow \mathbb{C}^{\times}$. For $f \in C_c(G)$, write $f_s : A \longrightarrow \mathbb{C}$, $a \longmapsto f(sa)$ and consider its Fourier transform

$$f_\chi(s) = \hat{f}_s(\chi) = \int_A f(sa)\,\overline{\chi(a)}\,da \quad .$$

One can check without difficulty that

(19.1) $$f_\chi(sa) = f_\chi(s)\,\chi(a) \quad , \quad |f_\chi| \in C_c(G/A) \quad .$$

On the other hand, the Plancherel formula on A gives

$$\|f_s\|^2 = \|\hat{f}_s\|^2$$

hence

$$\int_A |f(sa)|^2\,da = \int_{\hat{A}} |f_\chi(s)|^2\,d\chi$$

($d\chi$ is a suitably normalized Haar measure of the group $\hat{A}$) . Integrating the preceding equality on G/A

(19.2) $$\|f\|^2 = \int_G |f(s)|^2\,ds = \int_{\hat{A}} d\chi \int_{G/A} d\dot{s}\,|f_\chi(s)|^2 \quad .$$

On the space of continuous functions on G satisfying (19.1), we can define the norm

$$\|f_\chi\|^2 = \int_{G/A} d\dot{s}\,|f_\chi(s)|^2$$

and obtain the Hilbert space $L^2(G,\chi)$ by completion of this space $C_c(G,\chi)$. With these notations, we have obtained a decomposition

$$C_c(G) \longrightarrow \prod C_c(G,\chi) \quad : \quad f \longmapsto (f_\chi)$$

for which a "continuous Pythagoras theorem" is valid.

(19.3) $$\|f\|^2 = \int_{\hat{A}} \|f_\chi\|^2_{L^2(G,\chi)} \, d\chi .$$

We can say that the Hilbert space $L^2(G)$ has been disintegrated as a *Hilbert integral* over $\hat{A}$

(19.4) $$L^2(G) \xrightarrow{\sim} \int_{\hat{A}}^{\oplus} L^2(G,\chi)\, d\chi , \quad f \mapsto (f_\chi)_{\hat{A}} .$$

A better notation for (f_χ) would be $\int^{\oplus} f_\chi \, d\chi$ since the Fourier inversion formula for $f \in C_c(G)$ (hence $f_s \in C_c(A)$) reads

$$f_s(a) = \int_{\hat{A}} \hat{f}_s(\chi)\, \chi(a)\, d\chi .$$

(Recall that this inversion formula holds in quadratic mean: if one needs pointwise convergence of all integrals, one should take functions f of the form $f = f_1 * f_2$ with $f_i \in L^2(A)$.) For $a = e$, we indeed get

(19.5) $$f(s) = f_s(e) = \int_{\hat{A}} \hat{f}_s(\chi)\, d\chi = \int_{\hat{A}} f_\chi(s)\, d\chi .$$

All operations on functions have been effected on the right, hence commute with left translations. Taking $t^{-1}s$ instead of s in (19.5) we obtain

$$f(t^{-1}s) = \int f_\chi(t^{-1}s)\, d\chi$$

whence

$$\ell(t)f\,(s) = \int \ell_\chi(t) f_\chi\,(s)\, d\chi$$

$$\ell(t)f = \int \ell_\chi(t)\, f_\chi \, d\chi ,$$

or, even more succinctly

(19.6) $$\ell(t) = \int \ell_\chi(t)\, d\chi , \quad \ell = \int \ell_\chi \, d\chi .$$

This is a decomposition of the left regular representation, parametrized by $\hat{A}$. Unfortunately, even when A is a *maximal* abelian subgroup of G, the representations $(\ell_\chi, L^2(G,\chi))$ are not necessarily irreducible (even when G is a finite group!). Let us consider some examples. First

(19.7) $$G = SO_3(\mathbb{R}) \supset A = SO_2(\mathbb{R}) .$$

The irreducible representations of this group "are" real and $\pi \cong \pi^\vee$ so that we can write

$$L^2(G) \cong \bigoplus (V_\pi \otimes V_\pi) ,$$

$$L^2(G,\chi) \cong \bigoplus (V_\pi \otimes V_\pi^\chi)$$

where $V_\pi^\chi \subset V_\pi$ denotes the subspace (having dimension 0 or 1) of vectors which transform according to the character χ relatively to A. Thus we have

$$(\ell_\chi , L^2(G,\chi)) = \hat{\bigoplus}_{\pi|_A \supset \chi} V_\pi .$$

We see that this decomposition is given by the Frobenius reciprocity (sec.8, especially (8.10)) : $L^2(G,\chi)$ is the space of the induced representation $\mathrm{Ind}_A^G(\chi)$ and the decomposition of this representation "uses" all π such that $\pi|_A \supset \chi$ (and in fact all π except a finite number of them by sec.10).

(19.8) For G = $Sl_2(\mathbb{R})$, the situation is no better. One can take A = $SO_2(\mathbb{R})$ whence a "discrete" decomposition of $L^2(G)$ parametrized by the dual $\mathbb{Z}$ of A (the summands of this Hilbert sum must be reducible). Alternatively, one can take A = diagonal subgroup with positive entries

$$A \ni \begin{pmatrix} e^t & 0 \\ 0 & e^{-t} \end{pmatrix} = \mathrm{diag}(e^t, e^{-t}) \qquad (t \in \mathbb{R}) .$$

The dual $\hat{A}$ of A is isomorphic to $\mathbb{R}$ by identification of the unitary character

$$\mathrm{diag}(e^t, e^{-t}) \mapsto e^{its}$$

to the real number s. In this way, one gets a "continuous decomposition" of $L^2(G)$. But for this group, a "decomposition into irreducible components" of $L^2(G)$ has both a discrete part (sec.17) and a continuous part (sec.18, cf. Lang 1975).

Thus it comes nearly as a surprise that for certain particular groups, the preceding naîve decomposition gives a decomposition into irreducible ones. We quote two such examples.

(19.9) Let first G be the free group F_2 with two generators a and b. This group is a discrete group hence is unimodular. We take

$$A = \Gamma_a = \{a^n : n \in \mathbb{Z}\} .$$

The unitary characters of A are identified with complex numbers of modulus 1

$$\theta = \chi(a), \quad |\theta| = 1 \quad \text{satisfies} \quad \chi(a^n) = \theta^n .$$

Thus, the dual $\hat{A}$ of A is identified to the unit circle group in $\mathbb{C}$.
The decomposition of the left regular representation $(\ell, L^2(G))$ along A uses the spaces

$$H_\theta = \ell^2(G, \theta) \qquad (\theta \in U(1) = \{\theta \in \mathbb{C} : |\theta| = 1\})$$

consisting of functions f over the discrete space G satisfying

$$f(\gamma a^n) = f(\gamma)\,\theta^n \quad (n \in \mathbb{Z}) , \quad |f| \in \ell^2(G/\Gamma_a) .$$

An example of such a function is

$$\varepsilon(\gamma) = 0 \text{ if } \gamma \notin \Gamma_a , \quad \varepsilon(a^n) = \theta^n .$$

In fact, the left translates of this function ε with respect to a system of coset representatives of G/Γ_a make up an orthonormal basis of H_θ. We intend to show that in the disintegration

$$\ell^2(G) = \int_{|\theta|=1}^{\oplus} H_\theta \, d\theta , \quad \ell = \int_{|\theta|=1}^{\oplus} \ell_\theta \, d\theta ,$$

all representations (ℓ_θ, H_θ) are irreducible and inequivalent. To prove this contention, let us consider the operator $\alpha = \ell_\theta(a)$ having the function $\varepsilon = \varepsilon_\theta$ as eigenvector

$$\alpha(\varepsilon)(a^n) = \varepsilon(a^{n-1}) = \theta^{n-1} = \bar{\theta}\varepsilon(a^n) ,$$

$$\alpha(\varepsilon) \text{ vanishes outside } \Gamma_a ,$$

hence $\alpha(\varepsilon) = \bar{\theta}\cdot\varepsilon$. We even see that the multiples of ε are the only eigenvectors of α : if $f \in H_\theta$ satisfies $\alpha(f) = \lambda f$, take $\gamma \notin \Gamma_a$ and observe that the classes $a^n\gamma\Gamma_a$ $(n \in \mathbb{Z})$ are disjoint (if $a^n\gamma\Gamma_a$ has some elements in common with $a^m\gamma\Gamma_a$, there must exist an integer ℓ with $a^{n-m}\gamma a^\ell = \gamma$ hence n = m and $\ell = 0$). As $f \to 0$ at infinity on the discrete space G/Γ_a, we must have

$$\lambda^n f(\gamma) = \alpha^n f(\gamma) = f(a^{-n}\gamma) \to 0 \quad \text{when } |n| \to \infty .$$

As $\lambda \neq 0$, this implies $f(\gamma) = 0$ and f proportional to ε. Let us now take an operator T commuting with the representation ℓ_θ. Then $T(\varepsilon)$ must also be an eigenvector of α (with respect to the same eigenvalue $\bar{\theta}$)

$$\alpha(T\varepsilon) = T(\alpha\varepsilon) = T(\bar{\theta}\,\varepsilon) = \bar{\theta}\,T\varepsilon .$$

Hence $T\varepsilon = \lambda\varepsilon$ by the preceding observation. Then,

$$T(\ell_\theta(\gamma)\varepsilon) = \ell_\theta(\gamma)T(\varepsilon) = \lambda\,\ell_\theta(\gamma)\,\varepsilon .$$

But the elements $\ell_\theta(\gamma)\varepsilon$ form an orthonormal basis of H_θ when γ runs through a system of coset representatives of G/Γ_a, so that we deduce that $T = \lambda 1$ is a scalar operator. In particular, the only projectors commuting to ℓ_θ are 0 and 1 and this representation is irreducible.

Moreover, since an equivalence $T : \ell_\theta \xrightarrow{\sim} \ell_{\theta'}$ carries eigenvectors of $\ell_\theta(a)$ on eigenvectors of $\ell_{\theta'}(a)$ (with respect to the same eigenvalue), we infer that such an equivalence can exist only if $\theta' = \theta$.

A construction similar to the preceding one can of course be made with any generator of G in place of a, for example with b. We obtain a new disintegration

$$\ell^2(G) = \int_{|\zeta| = 1} K_\zeta \, d\zeta \quad , \quad \ell = \int_{|\zeta| = 1} \lambda_\zeta \, d\zeta \quad .$$

As before, the representations (λ_ζ, K_ζ) are irreducible and inequivalent. Moreover, all λ_ζ and ℓ_θ are inequivalent ! To see this, we show that the operator $\beta = \ell_\theta(b)$ has no eigenvector (whereas $\lambda_\zeta(b)$ has an eigenvector corresponding to the eigenvalue $\bar{\zeta}$). From an equality $\beta(f) = \mu \cdot f$, we would infer $f(b^{-n}\gamma) = \mu^n f(\gamma)$. But, as the cosets $b^n \gamma \Gamma_a$ $(n \in \mathbb{Z})$ are disjoint for every $\gamma \in G$, we should have

$$\mu^n f(\gamma) \longrightarrow 0 \quad \text{for} \quad |n| \longrightarrow \infty$$

and this again implies $f(\gamma) = 0 : f = 0$. As this example shows, two distinct disintegrations can be totally unrelated. The problem of decomposing the left regular representation of a locally compact group does not have a unique solution (not even unique up to isomorphism) and we shall have to introduce a class of groups for which it is well stated.

(19.10) As a second example, let F be a locally compact field and G the group consisting of matrices

$$\begin{pmatrix} a & b \\ 0 & 1 \end{pmatrix} \qquad (a \in F^\times , \ b \in F)$$

together with the subgroup A for which a = 1 (unipotent subgroup of G). This group G is not unimodular, but the restriction to A of its modular function is 1 (thus coïncides with the modular function of A since this group is commutative). For any unitary character $\chi \in \hat{F}$ we can consider the space of functions on G satisfying

$$f\left(\begin{pmatrix} 1 & b \\ 0 & 1 \end{pmatrix} g\right) = \chi(b) \, f(g) \quad . \tag{19.11}$$

To simplify notations, let us write f(a) instead of $f\begin{pmatrix} a & 0 \\ 0 & 1 \end{pmatrix}$ so that

$$f(a) = f\left(\begin{pmatrix} 1 & -b \\ 0 & 1 \end{pmatrix}\begin{pmatrix} a & b \\ 0 & 1 \end{pmatrix}\right) = \chi(b)^{-1} \, f\begin{pmatrix} a & b \\ 0 & 1 \end{pmatrix} .$$

Thus we have

$$f\begin{pmatrix} a & b \\ 0 & 1 \end{pmatrix} = \chi(b) \, f(a) \qquad (a \in F^\times).$$

In particular the function f on the group G has an absolute value which is a function on $G/A \cong F^\times$. The condition

$$(19.12) \qquad |f| \in L^2(G/A) \cong L^2(F^\times, d^\times a) = L^2(F^\times, |a|^{-1}\, da)$$

simply means

$$|a|^{-1} |f|^2 \in L^1(F^\times, da) = L^1(F)$$

or

$$|a|^{-\frac{1}{2}} |f| \in L^2(F) \quad .$$

This suggests to put $\varphi = |a|^{-\frac{1}{2}} f$ and to consider the representation π_χ in $L^2(F)$ which comes from the right regular representation in the space $L^2(G,\chi)$ defined by (19.11) and (19.12). One checks easily that this representation is given by the formula

$$(19.13) \qquad \pi_\chi\begin{pmatrix} a & b \\ 0 & 1 \end{pmatrix} \varphi(x) = |a|^{\frac{1}{2}} \chi_b(x)\, \varphi(xa) \qquad (\varphi \in L^2(F)) .$$

On this expression, the unitarity of the operators $\pi_\chi\begin{pmatrix} a & 0 \\ 0 & 1 \end{pmatrix}$ $(a \in F^\times)$ and $\pi_\chi\begin{pmatrix} 1 & b \\ 0 & 1 \end{pmatrix}$ $(b \in F)$ is obvious. I claim that <u>all these representations π_χ are irreducible provided</u> F <u>is not discrete and</u> $\chi \neq 1$. To follow the proof of this assertion, the reader who is not familiar with the theory of locally compact non-discrete fields (i.e. local fields) will have to assume $F = \mathbb{C}$ in which case the result is fully significant. First we have to recall that if $\chi \neq 1$, any unitary character of $F^\times$ is of the form χ_b (recall $\chi_b(x) = \chi(bx)$). Thus any operator T in $L^2(G,\chi)$ which commutes to all operators $\pi_\chi\begin{pmatrix} 1 & b \\ 0 & 1 \end{pmatrix}$ $(b \in F)$ has to be a multiplication operator by a bounded measurable function (as in the proof of (18.4)). If the operator T still commutes with the $\pi_\chi\begin{pmatrix} a & 0 \\ 0 & 1 \end{pmatrix}$ this bounded measurable function must still be invariant under all dilations $x \mapsto xa$ $(a \in F^\times)$ and thus be constant. This again proves that the only projectors commuting to π_χ are 0 and 1 and π_χ is irreducible.

Observe that this solvable group G is not unimodular, but trivial modifications will give similar results for the unimodular subgroup $G^1 = \operatorname{Ker} \Delta_G$ consisting of the matrices $\begin{pmatrix} a & b \\ 0 & 1 \end{pmatrix}$ with $a \in F^\times$, $|a| = 1$ and $b \in F$.

Appendix to sec. 19 : Note on Hilbertian integrals.

Let X be a locally compact space, m a positive measure on X and $x \longmapsto H_x$ a family of Hilbert spaces on X. A measurable structure (with respect to m) on this family consists of the data of a vector subspace $\mathcal{M}$ of the product $\prod_{x \in X} H_x$ satisfying

1) $x \longmapsto \|f(x)\|_{H_x}$ is measurable on X for each f ,
2) (saturation condition) if $x \longmapsto (g(x)|f(x))_{H_x}$ is measurable on X for all $g \in \mathcal{M}$, then $f \in \mathcal{M}$,
3) there is a sequence $(f_n)_{n \in \mathbb{N}} \subset \mathcal{M}$ with $\{f_n(x) : n \in \mathbb{N}\}$ total in H_x for each $x \in X$.

The elements $f \in \mathcal{M}$ are called *measurable fields*. The Hilbertian integral (with respect to m and $\mathcal{M}$)

$$\int_X^{\oplus} H_x \, dm(x)$$

consists of the measurable fields f for which $\int_X \|f(x)\|_{H_x}^2 \, dm(x) < \infty$.

This Hilbertian integral is a Hilbert space with respect to the scalar product

$$(f \mid g) = \int_X (f(x) \mid g(x))_{H_x} \, dm(x) \quad .$$

As usual, strictly speaking, the elements of the Hilbertian integral are *classes* of measurable fields (equal m-nearly everywhere...).

Practically, it may be better to start with a sequence of fields $(f_n)_{n \in \mathbb{N}}$ with $x \longmapsto \|f_n(x)\|_{H_x}$ measurable on X for all $n \in \mathbb{N}$ and $\{f_n(x) : n \in \mathbb{N}\}$ total in H_x for each $x \in X$. Then one defines the measurable fields f by

$$f \in \mathcal{M} \iff x \longmapsto (f_n(x) \mid f(x))_{H_x} \text{ measurable for all } n \in \mathbb{N}.$$

Then $\mathcal{M}$ is a measurable structure on the family $\prod H_x$ (only the first condition needs verification).

The elements f of the Hilbertian integral are also denoted by

$$f = \int_X^{\oplus} f(x) \, dm(x) \in \int_X^{\oplus} H_x \, dm(x) \quad .$$

Their "continuous" components satisfy a "continuous Pythagoras theorem" :

$$\|f\|^2 = \int_X \|f(x)\|_{H_x}^2 \, dm(x) \quad .$$

An interested reader who is familiar with the properties of measurable functions will be able to show that

$$\int_X^{\oplus} H_x \, dm(x) \neq \{0\}$$

provided the subset of X defined by $H_x \neq \{0\}$ is not m-negligible.

20 TYPE I GROUPS

When a representation ρ of a group G is an orthogonal (Hilbert) sum of representations all equivalent to some irreducible representation π of G, we say that ρ is *isotypical of type* π (or equivalently that ρ is a *multiple of* π). These representations can be written in the form

$$\rho = \pi \otimes 1 \quad \text{in} \quad H_\pi \otimes H$$

for some Hilbert space H. Taking an orthonormal basis of H, the matrices of the operators $\rho(x)$ take a diagonal form with equal blocks on the diagonal

$$\rho(x) = \begin{pmatrix} \pi(x) & & & 0 \\ & \pi(x) & & \\ & & \pi(x) & \\ 0 & & & \ddots \end{pmatrix} = \widehat{\bigoplus} \pi(x) \quad .$$

The algebra $\rho(G)'$ of operators commuting to all operators of $\rho(G)$ is

$$(\pi(G) \otimes 1)' = 1 \otimes \mathrm{End}(H)$$

(cf. ex.1 of sec. 7) and the operators which commute to $\rho(G)$ and have "the same form" as the $\rho(x)$ are scalar operators. More precisely, the operators having the same form as the $\rho(x)$ $(x \in G)$ are the operators in the von Neumann algebra generated by $\rho(G)$

$$\rho(G)'' = \overline{\langle \rho(G) \rangle}^{wk} \text{ (weak closure of linear combinations of operators } \rho(x),\ x \in G).$$

This von Neumann algebra is simply

$$\rho(G)'' = (1 \otimes \mathrm{End}(H))' = \mathrm{End}(H_\pi) \otimes 1$$

and the elements of this algebra which commute to $\rho(G)$ are the elements of the center of $\rho(G)'$ or of the center of $\rho(G)''$:

$$\rho(G)' \cap \rho(G)'' = (1 \otimes \mathrm{End}(H)) \cap (\mathrm{End}(H_\pi) \otimes 1) = \mathbb{C}\,1 \quad .$$

We give a definition.

(20.1) Definition. A unitary representation ρ of a group G is called factor representation when the von Neumann algebra generated by $\rho(G)$ has a center reduced to the scalar operators :

$$\rho(G)'' \cap \rho(G)' = \mathbb{C}\,1 \quad .$$

In the terminology of von Neumann algebras, ρ is a factor representation when the von Neumann algebra generated by $\rho(G)$ is a factor. By definition, isotypical representations of type π (for some irreducible representation π) are factor representations. In this case, the factor generated by $\rho(G)$ has minimal projectors $\neq 0$ and this property characterizes factors of *type I*.

(20.2) Definition. A group of type I is a locally compact group G for which all factor representations are isotypical (all factor representations generate a factor of type I).

(20.3) Example. Let G be a discrete group with infinite conjugation classes $\neq \{e\}$ (e.g. $G = F_2$ free group on two generators as in (19.9)). The left regular representation $(\ell, \ell^2(G))$ is a factor representation. Let indeed

$$\mathcal{U} = \mathcal{U}(G) = \ell(G)'' = \overline{\langle \ell(G) \rangle}^{wk}$$

be the von Neumann algebra generated by the left translation operators. Any operator $T \in \mathcal{U}$ is given by a convolution operator

$$T(\varphi) = f_T * \varphi$$

where

$$f_T = T(\varepsilon_e) \in \ell^2(G) ,$$

ε_e = characteristic function of the neutral element $e \in G$.

If we assume that T lies in the center of $\mathcal{U}$, we must have

$$f_T = \varepsilon_x * f_T * \varepsilon_{x^{-1}} \qquad (x \in G)$$

hence f_T must be constant on the conjugation classes of G. But $f_T \in \ell^2(G)$ implies that $f_T \to 0$ on the discrete space G. Thus f_T must vanish on all infinite conjugation classes of G. In our case, we see that $f_T = c\,\varepsilon_e$ must be a multiple of ε_e and $T = c\cdot 1$ is a scalar operator.

It can be shown that these factors $\mathcal{U}(G)$ are of type II_1 .
Hence, these groups are not of type I.

Here is a general construction of factor representations.

(20.4) Proposition. Let $G = G_1 \times G_2$ be the product of two locally compact groups and π a unitary irreducible representation of G. Then the restriction of π to G_1 is a factor representation.

Proof. We have to show that if $T \in \pi(G_1)''$ commutes with $\pi(G_1)$, then T is a scalar operator. But $\pi(G_1)$ commutes to $\pi(G_2)$ hence

$$\pi(G_1)'' \text{ still commutes to } \pi(G_2)$$

(by the von Neumann density theorem). Thus T commutes with $\pi(G_1)$ and $\pi(G_2)$: T commutes with $\pi(G)$. Schur's lemma implies that T is a scalar operator. q.e.d.

(20.5) Remark. The conclusion of the preceding proposition still holds if we only assume the original representation π of $G = G_1 \times G_2$ to be a factor representation. Indeed, $T \in \pi(G_1)'' \subset \pi(G)''$ shows that T is in the center of the von Neumann algebra generated by $\pi(G)$.

The following lemma is to be compared with (14.10).

(20.6) Lemma. Let π be a factor representation of a locally compact group G. If there is a compactly supported continuous function $f \in C_c(G)$ with $0 \neq \pi(f)$ compact operator, then π is a multiple of an irreducible representation (it generates a factor of type I).

Proof. The operator $\pi(f)$ belongs to the von Neumann algebra $\pi(G)''$ generated by $\pi(G)$. The same is true of $\pi(f)^*$ and hence of

$$\pi(f)^*\pi(f) = \pi(f^* * f) \qquad \text{(cf. (14.2))} .$$

This operator is hermitian compact and non-zero

$$\pi(f)x \neq 0 \implies \|\pi(f)x\|^2 = (x \mid \pi(f)^*\pi(f)x) \neq 0 .$$

Let $\lambda > 0$ be an eigenvalue of this operator and P_λ be the spectral projector (with finite rank) corresponding to λ :

$$P_\lambda \in \{\pi(f)^*\pi(f)\}'' \subset \pi(G)'' .$$

As P_λ has a finite dimensional image, $\pi(G)''$ has some minimal projectors and is a type I von Neumann algebra.

(20.7) Proposition. *Let $G = G_1 \times G_2$ be the product of two locally compact groups and π a factor representation of G such that the restriction of π to G_1 is isotypical (for example, assume G_1 of type I) :*

$$\pi|_{G_1} \cong \pi_1 \otimes 1 \quad \text{with } \pi_1 \text{ irreducible representation of } G_1 .$$

Then there is a factor representation π_2 of G_2 and an equivalence

$$\pi \cong \pi_1 \otimes \pi_2 \quad .$$

If π is irreducible, π_2 is also irreducible.

Proof. By assumption, the operators belonging to $\pi(G_1) \subset \pi(G_1)''$ can be written in the form $\pi_1(x_1) \otimes 1$ in a tensor product space $H_1 \hat{\otimes} H_2$. As π_1 is assumed to be irreducible, these operators generate the von Neumann algebra $\mathrm{End}(H_1) \otimes 1$. The operators $\pi(x_2)$ $(x_2 \in G_2 \subset G_1 \times G_2)$ must commute to all preceding operators hence can be written in the form $1 \otimes \pi_2(x_2)$. This gives a construction of a representation π_2 of G_2 . We have to show that this representation π_2 is a factor representation (irreducible if π is irreducible). If $T \in \pi_2(G_2)'' \subset \mathrm{End}(H_2)$ commutes with $\pi_2(G_2)$, $1 \otimes T$ will commute with $\pi(G_1) = \pi_1(G_1) \otimes 1$ and with $1 \otimes \pi_2(G_2)$ hence with $\pi(G)$. But this operator belongs to $1 \otimes \pi_2(G_2)'' \subset$ $\subset \pi(G)''$ hence is a scalar operator. The proposition follows .

(20.8) Proposition. *The product $G = G_1 \times G_2$ of two groups of type I is a group of type I.*

Proof. Take a factor representation π of $G = G_1 \times G_2$. We have seen that $\pi|_{G_1}$ is a factor representation (20.4). As we are assuming that G_1 is a group of type I, this factor representation is isotypical, i.e.

$$\pi|_{G_1} \quad \text{equivalent to} \quad \pi_1 \otimes 1$$

where π_1 is some unitary irreducible representation of G_1. By (20.7), we can find a factor representation ρ of G_2 such that $\pi \cong \pi_1 \otimes \rho$. Since we also assume that G_2 is a type I group, this factor representation ρ is isotypical, say $\rho \cong \pi_2 \otimes 1$ with some irreducible representation π_2 of G_2 . Thus we have

$$\pi \cong \pi_1 \otimes \pi_2 \otimes 1 \cong \pi_0 \otimes 1 \quad .$$

Obviously $\pi_0 = \pi_1 \otimes \pi_2$ is irreducible ($\pi_1 \otimes \pi_2(G) = \pi_1(G_1) \otimes \pi_2(G_2)$ generates the von Neumann algebra $\mathrm{End}(H_1 \otimes H_2)$) and the representation $\pi \cong \pi_0 \otimes 1$ is isotypical. q.e.d.

When working with factor representations, it is sometimes convenient to identify the representations (π, H_π) and $(\pi \otimes 1, H_\pi \otimes H)$, thus identifying in particular all isotypical representations of the same type. A way of achieving this identification is to introduce the following equivalence relation.

(20.9) Definition. Two factor representations π, σ of a group G are called *quasi-equivalent* when the mapping $\pi(x) \mapsto \sigma(x)$ (defined on $\pi(G) \subset \mathrm{End}(H_\pi)$) extends to an isomorphism of the factors generated by $\pi(G)$ and $\sigma(G)$ respectively. The set of quasi-equivalence classes of factor representations of G is called *quasi-dual* $\overset{\frown}{G}$ of G.

Of course, each unitary irreducible representation of G defines a quasi-equivalence class and two equivalent unitary irreducible representations of G are quasi-equivalent. Thus we get a canonical map

$$\hat{G} \longrightarrow \overset{\frown}{G}$$

from the unitary dual of G (set of equivalence classes of unitary irreducible representations of G) to the quasi-dual of G. When the group G is of type I, this map is bijective by definition and the two sets can be identified.

To be able to give a general decomposition theorem into factor representations, it is necessary to define a suitable Borel structure on the set $\overset{\frown}{G}$. Without going into details and proofs, let us just indicate how this can be done (details can be found in Dixmier 1964, where other references can also be found).

First, we have to assume that G is separable (i.e. has a countable basis for the open sets). Let $H_n = \ell^2([0,n[)$ denote the standard Hilbert space of dimension $n \leqslant \infty$ (for $n = \infty$, we thus take $H_\infty = \ell^2([0,\infty[) = \ell^2(\mathbb{N})$). On the set $\mathrm{Fac}_n(G)$ of factor representations of G in H_n, there is a topology for which $\pi_i \longrightarrow \pi$ precisely when

$$\forall\, v \in H_n\ ,\ \pi_i(.)v \longrightarrow \pi(.)v \text{ on all compact sets of } G\ .$$

On the disjoint union Fac(G) of all $\mathrm{Fac}_n(G)$ $(n \leqslant \infty)$, we put the sum topology. The Borel structure of the quasi-dual of G is defined by taking the image structure under the canonical map

$$\mathrm{Fac}(G) \longrightarrow \overset{\frown}{G}\ .$$

(20.10) Theorem, Let π be a unitary representation of a separable group in a separable space H. Then there is a positive measure m on the quasi-dual $X = \widehat{G}$ which realizes a decomposition of π into factor representations in the following sense. There is an isomorphism $H \xrightarrow{\sim} \int_X^{\oplus} H_s \, dm(s)$ leading to an equivalence $\pi \xrightarrow{\sim} \int_X^{\oplus} \pi_s \, dm(s)$ where each (π_s, H_s) is in the class of $s \in X = \widehat{G}$.

Moreover, the measure $m = m_\pi$ is essentially uniquely determined by π. Let us recall that two measures m and m' on X are called *equivalent* when they have same negligible sets (thus there exists a nowhere vanishing measurable function f with $m' = f \cdot m$). Then, under the conditions of (20.10), the class of m is well determined. This measure m is obtained by disintegration of the commutative von Neumann algebra $\mathfrak{Z}$ = center of the von Neumann algebra generated by $\pi(G)$ ($= \pi(G)' \cap \pi(G)''$).

As the example (19.9) has shown, there is no uniqueness for finer decompositions into irreducible components, in general (cf. also (20.3)).

(20.11) Some comments. The theorem (20.10) can in particular be applied to the left regular representation of any (separable, unimodular) locally compact group G, even when G is not of type I. For example, it can be applied to the groups

$$G = G_F = \left\{ \begin{pmatrix} a & b \\ 0 & 1 \end{pmatrix} : a \in F^\times, b \in F \right\}$$

where F is an infinite discrete (countable) field. These groups are discrete, hence locally compact and unimodular. But one can show that their left regular representation can be disintegrated into two non-equivalent Hilbert sums of irreducible ones (one can "induce" either from the diagonal subgroup or from the unipotent subgroup defined by $a = 1$ to get such disintegrations). In the next section, we shall consider this situation for groups of type I.

EXERCISES

1. *Let* π *be a factor representation of a group* G *in some Hilbert space* H. *Let* $H_1 \subset H$ *be a closed invariant subspace of* H *(with respect to this representation). Show that the induced representation* $\pi_1 : G \rightarrow Gl(H_1)$ *is a factor representation (in* H_1*).*

2. *Let* G *and* H *be two locally compact groups,* $\pi \in \hat{G}$ *and* $\pi_1 , \pi_2 \in \hat{H}$ *be irreducible representations such that*

$$\pi \otimes \pi_1 \quad \text{and} \quad \pi \otimes \pi_2 \quad \text{are equivalent}$$

(as representations of $G \times H$*). Show that* π_1 *is equivalent to* π_2 *.*

21 GETTING NEAR AN ABSTRACT PLANCHEREL FORMULA

In this section, G will always denote a locally compact *unimodular* group. When necessary, we shall even assume that the topology of G has a countable basis for the open sets (in this case, G is *metrizable* and has a *countable dense* subset).

As in the preceding section, we motivate a definition by the consideration of a finite dimensional situation. Let π be a representation of some group G which is a finite sum of inequivalent irreducible representations. In a suitable basis of the representation space, the matrix form of π is given by diagonal blocks (of varied sizes) and Schur's lemma shows that if an operator commutes to π, it is given by a scalar multiplication in each irreducible subspace. In particular, the algebra of operators commuting with $\pi(G)$ is a commutative algebra : it is a subalgebra of the algebra of diagonal operators in any basis compatible with the decomposition of π into irreducible components. Conversely, if $\pi: G \longrightarrow Gl(V)$ is a finite-dimensional representation with a commutative commuting algebra $\pi(G)' \subset End(V)$, it is often possible to diagonalize this algebra by choosing a suitable basis of the representation space. This will be the case if $\pi(G)'$ is stable under the operation of taking the adjoint (e.g. π unitary). In this way, we shall find a decomposition of π as a sum of inequivalent irreducible subrepresentations. In general, we give a definition.

(21.1) Definition. We say that a unitary representation $\pi: G \longrightarrow Gl(H)$ has a *simple spectrum* when its commuting algebra $\pi(G)' \subset End(H)$ is commutative.

By definition, π has a simple spectrum exactly when

$$\pi(G)' \subset \pi(G)'' ,$$

i.e. $\pi(G)'$ = center of the von Neumann algebra generated by $\pi(G)$. As an immediate consequence of the definition, we note that

(21.2) a factor representation with simple spectrum is irreducible.

To be able to prove that the biregular representation of a unimodular group G has a simple spectrum, we need the commutation theorem of Godement-Segal ((21.9) below).

We first construct the *Hilbert algebra* of the group G. Let $H = L^2(G)$ throughout this section.

(21.3) Definition. A *moderate* function f on G is an element $f \in H$ such that the operator $g \mapsto f * g : C_c(G) \to H$ extends continuously to H.

The treatment given in sec.14 shows that all $f \in L^1(G) \cap H$ are moderate . Let us denote by J the unitary operator in H given by

$$J(f) = f^* \quad (\text{where } f^*(x) = \overline{f(x^{-1})}) .$$

Obviously

$$J(f * g) = Jg * Jf$$

and a function f is moderate precisely when $g \mapsto g * f$ (defined on $C_c(G)$) extends continuously to H. For f and $g \in C_c(G)$ first, let us define

$$(21.4) \qquad U_f(g) = f * g = V_g(f) .$$

(21.5) Lemma. When f is moderate and $g \in H$, one still has $U_f(g) = f * g$ (where U_f still denotes the continuous extension of the precedingly defined operator $C_c(G) \to H$) .

Proof. a) Take first $g \in C_c(G)$. There is a sequence $(f_n) \subset C_c(G)$ which converges to f in H. Thus

$$f_n * g \to f * g \quad \text{in } L^2(G) = H$$

and

$$f_n * g = V_g(f_n) \to V_g(f) = U_f(g)$$

proves the assertion in this particular case.

b) Take now $g \in H$ arbitrary. There exists a sequence $(g_n) \subset C_c(G)$ with $g_n \to g$ in H . Thus

$$f * g_n \underset{a)}{=} U_f(g_n) \to U_f(g) \quad \text{in } H .$$

But the Cauchy-Schwarz inequality shows that

$$|f * g(x) - f * g_n(x)| \leqslant \int |f(y)(g - g_n)(y^{-1}x)|\, dy \leqslant$$

$$\leqslant \|f\| \, \|g - g_n\| \quad (\text{in } H : \text{quadratic norms ! }) ,$$

so that $f * g_n \to f * g$ *uniformly* on G, whence the conclusion.

We recall that

$$(f \mid g) = f^* * g\,(e) \qquad (f, g \in H).$$

Thus, the equality

$$(f * g \mid h) = (g \mid f^* * h)$$

is an immediate consequence of the associativity of the convolution :

$$(f * g)^* * h = (g^* * f^*) * h = g^* * (f^* * h)$$

(when all products are defined...)

(21.6) Definition. The Hilbert algebra A = A(G) of the group G is the convolution algebra of moderate functions on G : If $f \in A$, then $f^* \in A$ and

$$(g^* \mid f^*) = (f \mid g) \ , \quad U_f^* = U_{f^*} \ , \quad V_g^* = V_{g^*}$$

for $f, g \in A$.

By definition, $C_c(G) \subset A \subset H$ so that A is dense in H .

Let us prove the following identity

(21.7) $$J\, U_{f^*}\, J = V_f \qquad (f \in A) .$$

For x and y in A first, we have

$$(J\, U_{f^*}\, J x \mid y) = (y^* \mid U_{f^*}\, J x) = (y^* \mid f^* * x^*) =$$
$$= (x * f \mid y) = (V_f x \mid y) .$$

The density of A in H together with the continuity of the operators $J\, U_{f^*}\, J$, V_f gives the equality (21.7).

(21.8) Theorem (Godement). Let us denote by $\mathcal{U}$ (resp. $\mathcal{V}$) the von Neumann algebra in End(H) generated by left (resp. right) translations of G. Then the set

$$I = \{ U_f : f \text{ moderate in } H \}$$

is a $*$-ideal, weakly dense in $\mathcal{V}'$.

Proof. For $x, y \in A$, we have

$$U_f V_x y = U_f(y * x) = V_x V_y f = V_x U_f y ,$$

hence

$$U_f V_x = V_x U_f \in \mathrm{End}(H) \quad \text{if } f \in A .$$

This proves that U_f commutes to $\mathcal{V}$ when f is moderate (an operator commutes to $\mathcal{V}$ as soon as it commutes to the right convolutions with elements of $C_c(G)$):

$$U_f \in \mathcal{V}' \text{ when f is moderate.}$$

Moreover, if $T \in \mathcal{V}'$,

$$T U_f x = T V_x f = V_x T f \text{ is continuous in } x ,$$

proves that

$$T f \text{ is moderate and } U_{Tf} = T U_f \in I .$$

Since $U_f^* = U_{f^*}$, I is a $*$-ideal of $\mathcal{V}'$. Since composition of operators is separately weakly continuous, $\mathcal{V}'$ is weakly closed and the proof will be complete if we show that $\mathcal{V}'$ is the weak closure I" of I. It is enough to show that $\mathcal{V}' \subset I''$. Thus, we take any $T \in \mathcal{V}'$, $X \in I'$ and show TX = XT. But we know that for each $x \in A$, $T U_x \in I$. Hence

$$T U_x \text{ commutes with } X : \quad T U_x X = X T U_x .$$

Taking a sequence x_n such that $U_{x_n} \to 1_H$ (for the strong topology), we conclude $T X = X T$ as claimed. q.e.d.

(21.9) <u>Corollary</u> (<u>Godement-Segal</u>). <u>We have</u> $\mathcal{U}' = \mathcal{V}$, $\mathcal{V}' = \mathcal{U}$.

<u>Proof</u>. As $\mathcal{U}$ and $\mathcal{V}$ commute, $\mathcal{U} \subset \mathcal{V}'$ and by taking the weak closure $\mathcal{U}'' \subset \mathcal{V}'$. It will thus be enough to show $\mathcal{U}'' \supset \mathcal{V}'$, or by the above theorem (21.8), $\mathcal{U}'' \supset I$ i.e. $\mathcal{U}'' \ni U_f$ for all moderate functions $f \in H$. Thus we have to show that all elements U_f (f moderate) commute to $\mathcal{U}'$. Still by the above theorem with left and right exchanged (consider (21.8) for the opposite group...), it is enough to show

$$U_f \text{ commutes to all } V_g \text{ (g moderate)} \qquad \text{(f moderate).}$$

This follows from (21.5) and associativity of convolution. In detail, if h is moderate, $y = V_g h$ is also moderate so that $U_f y = V_y f$ and

$$\begin{aligned} U_f V_g h &= U_f y = V_y f = V_{V_g h} f = U_f(V_g h) = \\ &= V_g V_h f \qquad \text{(since } V_g \in \mathcal{U}'\text{)} \\ &= V_g U_f h . \end{aligned}$$

This gives the conclusion.

(21.10) Corollary. When G is unimodular, the biregular representation of G in $H = L^2(G)$ is unitary and has a simple spectrum.

Proof. We have to show that the commuter of the von Neumann algebra generated by $\mathcal{U}$ and $\mathcal{V}$ is commutative. But this commuter is

$$(\mathcal{U} \cup \mathcal{V})' = \mathcal{U}' \cap \mathcal{V}' = \mathcal{V} \cap \mathcal{V}' = \\ = \mathcal{U}' \cap \mathcal{U} = \mathcal{V} \cap \mathcal{U} = \mathcal{Z}$$

is the common center of $\mathcal{U}$ and $\mathcal{V}$.

(21.11) When $G = \Gamma$ is a discrete group with infinite conjugation classes different from $\{e\}$ (group ICC as in (19.9)) its biregular representation is irreducible (cf. (20.3) and (21.2)).

There is a general decomposition theorem for simple spectrum representations (compare with (20.10)). We state it now.

(21.12) Theorem. Let G be a (separable) unimodular locally compact group and ρ a unitary representation of G in a separable Hilbert space H with simple spectrum. Then, there is a positive measure m on $\hat{G}$ and an equivalence

$$\rho \xrightarrow{\sim} \int_{\hat{G}}^{\oplus} \pi \, dm(\pi) \quad , \quad H \xrightarrow{\sim} \int_{\hat{G}}^{\oplus} H_\pi \, dm(\pi) \quad .$$

Again, we refer to Dixmier 1964 for a proof (observe that the decomposition is made according to the commutative algebra $\rho(G)'$).

Still assuming G to be (separable) and unimodular, we can apply the preceding theorem to the biregular representation of G in $L^2(G)$ (cf. (21.10)) and we get thus a positive measure m on $(G \times G)^\wedge$ decomposing this representation. When G is of type I,

$$(G \times G)^\wedge \cong \hat{G} \times \hat{G}$$

(cf. (20.7)) and we can write

$$\ell \times r \cong \int \pi \otimes \sigma \, dm(\pi,\sigma) \quad , \quad L^2(G) \cong \int H_\pi \hat{\otimes} H_\sigma \, dm(\pi,\sigma) \quad .$$

We can say more in this case. The identity

(21.13) $\quad J \circ \ell \times r\,(s,t) f \quad = \quad r \times \ell\,(s,t)\, Jf \qquad (f \in H = L^2(G))$

is immediately verified :

$$J.\ell\times r(s,t)\,f(x) = \overline{\ell\times r(s,t)f(x^{-1})} = \overline{f(s^{-1}x^{-1}t)} =$$

$$= f^*(t^{-1}xs) = r\times\ell\,(s,t)\,Jf(x) \quad .$$

Since J is anti-linear, it induces an equivalence

$$J : \quad \ell\times r \longrightarrow \overline{r\times\ell}$$

Moreover, the identity (21.7) $J\,U_{f*}\,J = V_f$ gives

$$(21.14) \qquad J\,c\,J = c^* \qquad (c\in\mathcal{Z})$$

(observe that $U_{f*} = (U_f)^*$ and $U_f = V_f$ iff $U_f \in \mathcal{Z}$). Thus we have two decompositions of the biregular representation of G in $L^2(G)$:

$$J : \quad \int \pi\otimes\sigma \; dm(\pi,\sigma) \;\tilde{\rightarrow}\; \int \bar{\sigma}\otimes\bar{\pi}\; dm(\pi,\sigma) \quad .$$

Since there is a relative uniqueness in the decomposition given by (21.12) (just as in (20.10)) , we infer that

$$\pi\otimes\sigma \cong \bar{\sigma}\otimes\bar{\pi}$$

nearly everywhere on the support of the measure m. Thus we can assume $\pi = \bar{\sigma}$ $(\sigma = \bar{\pi})$ on the support of m. The support of m is contained in the subset

$$\{\pi\otimes\bar{\pi} \;:\; \pi\in\hat{G}\} = \{\bar{\pi}\otimes\pi \;:\; \pi\in\hat{G}\}$$

and we can even write

$$\ell\times r = \int_{\hat{G}} \bar{\pi}\otimes\pi \; d\mu(\pi)$$

for a certain positive measure μ on $\hat{G}$. The class of the measure μ is well determined by the general theorems already alluded to : replacing μ by $f\mu$ simply amounts to replacing the scalar products

$$(.|.)_{H_\pi} \text{ by } f^{-1}(\pi)\,(.|.)_{H_\pi} \quad .$$

But since the space of $\bar{\pi}\otimes\pi$ can be canonically identified to $H_{\check{\pi}}\hat{\otimes}H_\pi$ (Hilbert space consisting of the Hilbert-Schmidt operators on H_π), it carries *a canonical scalar product*. Indeed, the scalar product of H_π is determined up to a positive constant by the requirement that π is unitary, and an amplification by a factor $\lambda > 0$ of the scalar product of H_π produces an amplification of a factor $1/\lambda$ of the scalar product of $H_{\check{\pi}}$, thus leaving invariant the scalar product of $H_\pi\hat{\otimes}H_\pi$ (corresponding to the Hilbert-Schmidt scalar product on operators).

For these canonical choices, the corresponding measure μ on $\hat{G}$ is the Plancherel measure μ_{Pl} . It is well determined up to a positive constant (and completely well determined if a Haar measure on G is fixed). By definition, we have

$$\ell \times r \cong \int_{\hat{G}} \bar{\pi} \otimes \pi \; d\mu_{Pl}(\pi) \quad ,$$

$$L^2(G) \cong \int_{\hat{G}}^{\oplus} \mathrm{End}_2(H_\pi) \; d\mu_{Pl}(\pi) \quad .$$

This last identity can be written

$$\|f\|^2 = \int_{\hat{G}} \|\pi(f)\|_2^2 \; d\mu_{Pl}(\pi) \qquad (f \in H = L^2(G))$$

Minimal closed invariant subspaces of this biregular representation are given by the points $\pi \in \hat{G}$ for which $\mu_{Pl}(\pi) > 0$. These points are precisely the discrete series representations and as we have seen in sec.16, $\mu_{Pl}(\pi) = d_\pi$ for these representations. The discrete series gives the *atomic part of the Plancherel measure*. The Plancherel formula for a type I group takes the general form

$$\|f\|^2 = \sum_{\Omega} d_\pi \|\pi(f)\|_2^2 + \int_{\hat{G}} \|\pi(f)\|_2^2 \; d\mu'_{Pl}(\pi)$$

with both discrete and continuous parts (we say that the measure μ'_{Pl} for which $\mu'_{Pl}(\pi) = 0$ for *all* $\pi \in \hat{G}$ is a *diffuse measure*). This formula generalizes the cases G compact and $G = \mathbb{R}$ (locally compact and abelian).

In general, the support of the Plancherel measure is strictly smaller than $\hat{G}$, and one defines the *reduced dual* of G by

$$\hat{G}_{red} = \mathrm{Supp}\, \mu_{Pl} \subset \hat{G} \quad .$$

For $G = Sl_2(\mathbb{R})$, one can show that the reduced dual consists of the discrete series (as in any locally compact unimodular group) and the principal series. But in this example $\hat{G}$ also contains a *supplementary series* (cf. Lang 1975).

One can also show that

$$\hat{G}_{red} = \{\pi \in \hat{G} : \mathrm{Ker}\, \pi^* \supset \mathrm{Ker}\, \ell^*\} = \{\pi \in \hat{G} : \mathrm{Ker}\, \pi^* \supset \mathrm{Ker}\, r^*\}$$

where $\rho \mapsto \rho^*$ denotes the extension of representations from G to $C^*(G)$. (Recall that the left and right regular representations are equivalent!)

EPILOGUE

The following analogue of the Peter-Weyl theorem (p.29) has not been proved.

<u>Gelfand-Raikov theorem</u>. <u>For any locally compact group</u> G <u>and any</u> $x \neq e \in G$, <u>there is an irreducible unitary representation</u> $\pi \in \hat{G}$ <u>such that</u> $\pi(x) \neq \mathrm{id}$.

However, this theorem follows easily from the deeper results (20.10) or (21.12) applied to the (bi-)regular representation of G. They show that one can even take π in the reduced dual $\hat{G}_{red}$. (For details, cf. Dixmier 1964 or Gaal 1973.)

On the other hand, our introduction of type I groups has been made in an ad hoc way. A proof that a certain class of groups (semi-simple real or p-adic algebraic groups) is of type I is not trivial and usually follows the following pattern.

a) Find a large (or maximal) compact subgroup K of G such that $\pi|_K$ has finite multiplicities for all $\pi \in \hat{G}$.

b) Prove that $\pi(f)$ is a compact operator for all $\pi \in \hat{G}$ and $f \in C_c(G)$ (or $f \in L^1(G)$ or even $f \in C^*(G)$) (groups with this property are called CCR groups).

c) Prove that CCR groups are type I .

Interested readers will find details in Dixmier 1964 : liminaire = CCR (= completely continuous representations), postliminaire = GCR $\Longrightarrow$ CCR .

REFERENCES

Adams J.F. LECTURES ON LIE GROUPS W.A. Benjamin, Inc. 1969

Borel A. REPRESENTATIONS DE GROUPES LOCALEMENT COMPACTS Lect. Notes in Math., 276, Springer-Verlag 1972

Bourbaki N. TOPOLOGIE GENERALE Chap. I à IV , Hermann Paris 1971

--- ESPACES VECTORIELS TOPOLOGIQUES Chap. I à V, Masson 1981

--- INTEGRATION Chap. I à IV, Hermann 1965

--- INTEGRATION Chap. VI , Hermann 1959

--- INTEGRATION Chap. VII et VIII, Hermann 1963

Chevalley C. THEORY OF LIE GROUPS I , Princeton Univ. Press 1946

Dieudonné J. FOUNDATIONS OF MODERN ANALYSIS Acad. Press 1960

--- ELEMENTS D'ANALYSE vol.2, Gauthier-Villars, Paris 1969

--- ELEMENTS D'ANALYSE vol.3, Gauthier-Villars, Paris 1970

--- ELEMENTS D'ANALYSE vol.5, Gauthier-Villars (Bordas) 1975

Dixmier J. LES C*-ALGEBRES ET LEURS REPRESENTATIONS Gauthier-Villars, Paris 1964

Fefferman C. PROC. INT. CONGRESS MATH. Vancouver 1974 vol.1 (95-118)

Gaal S.A. LINEAR ANALYSIS AND REPRESENTATION THEORY Springer-Verlag 1973

Lang. S. $SL_2(\mathbb{R})$ Addison-Wesley 1975

Riesz F., Nagy B., LEÇONS D'ANALYSE FONCTIONNELLE Gauthier-Villars, Paris 1975 (6^e ed.)

Robert A. SYSTEMES DE POLYNOMES Queen's Univ. 1973 (Math.ser. # 35)

--- EXEMPLES DE GROUPES DE FELL C.R.Acad.Sc.Paris 1978 t.287, pp. 603-606 .

Rudin W. FOURIER ANALYSIS ON GROUPS, Interscience Publ. 1962

--- FUNCTIONAL ANALYSIS McGraw-Hill Publ.Co. 1973

Serre J.-P. REPRESENTATIONS LINEAIRES DES GROUPES FINIS Hermann Paris 1967, 1971

Weil A. L'INTEGRATION DANS LES GROUPES TOPOLOGIQUES Hermann Paris 1953

INDEX

MAIN THEOREMS